HORIZONTAL SUBSURFACE DRAINAGE THEORIES FOR AGRICULTURAL LANDS

Chhedi Lal

Senior Scientist
Central Soil Salinity Research Institute
Regional Research Station, 21/467, II Floor,
ICCMRT Building, Ring Road, Indira Nagar,
Lucknow - 226 016 (UP), INDIA

INTERNATIONAL BOOK DISTRIBUTING CO.

(Publishing Division)

Published by
INTERNATIONAL BOOK DISTRIBUTING CO.
(Publishing Division)
Chaman Studio Building, 2nd Floor,
Charbagh, Lucknow 226 004 U.P. (INDIA)
Tel. : Off. : 2450004, 2450007, 2459058 Fax : 0522-2458629
E-Mail : ibdco@sancharnet.in

First Edition. 2004

Price: Rs. 550/-

ISBN 81-8189-027-2

Composed & Designed at :
Panacea Computers
33, Nehru Road,
Sadar Cantt. Lucknow-226 002
Phone : 2481164, 2483354, 3127082
E-mail : prasgupt@rediffmail.com

Printed at:
Army Printing Press
33, Nehru Road,
Sadar Cantt. Lucknow-226 002
Phone : 2481164, 2483354, 3246263
E-mail : armypress@indiatimes.com

FOREWORD

This book is intended to serve as a text or a reference book for engineering students studying at senior level in their college or university education. The main purpose of this book is to present theories, principles, and approaches to the design of drainage systems. The history of agricultural drainage design dates back to the Greek historical Herodotus in 400 B.C. with reference to the Nile Valley in ancient Egypt. Many Roman authors (Cato, Virgil, and others) also have given reference to agricultural and urban drainage design systems using channels in the early eastern nations. Depending on the climatic, hydrology, landscape, crops, soil types, and management factors, different types of drainage design methods have been used for different agricultural fields during the history of mankind. This book does an outstanding job in compiling all drainage design methods and associated drain design formulas. Several boundary value problems have been developed for different field conditions and the needed solutions were derived using mathematical models so that practicing engineers can use these mathematically derived solutions to design surface and subsurface drain systems.

This book addresses special drainage needs for the irrigated agricultural fields. Irrigated agriculture increases crop yields and economic health of farmers but it also could raise shallow groundwater tables and salinity levels because of excessive irrigations. Therefore, properly designed irrigation systems for agricultural crops must have adequately designed drainage systems. Next questions is : what should a successful drainage system have to promote its long term effectiveness or sustainability in the agro-ecosystem ? Depending on the climate, hydrogeology, soil characteristics, topography, farming practices, the drainage effluents and their effects on soil and water quality need to be managed. The disposal and management of drainage effluents from agricultural fields need special attention to minimize its offsite effects on the environment. This book will serve as a useful reference book to answer some of these questions.

This book has five different chapters, chapter 3 to be the largest and most comprehensive in material and thoughtfulness. The chapter 1 is an introductory chapter giving definitions of various types of drainage and needs of drainage for agricultural lands. This chapter sets the foundation for writing rest of the chapters of this book. The chapter 5 is written on "saturated hydraulic conductivity". This chapter begins with a definition of conductivity and then gives extensive and clear details on laboratory and in-situ field methods to measure saturated hydraulic conductivity. This chapter also describes theories behind these methods used for the determination of hydraulic conductivity. For each method several working examples are given for the benefit of the students and to bring a better understanding on the use of these methods to determine saturated hydraulic conductivity.

The chapter 3 gives steady-state and unsteady-state solutions for boundary value problems for field collector and main drains using different types of flow domain systems such as cylindrical, elliptical and spherical. Drainage theories by Dupuit-Forchheimer and Hooghoudt (1937) have been used to develop drainage models for sloppy lands and layered soil profiles. Also, theories developed by Khirkham and Grover have been used to determine drain spacing for soils with varying hydraulic conductivity. In addition this chapter provides solutions for steady-state and unsteady-state drainage systems under fluctuating water table conditions. In reality, this chapter provides methods for the design of drainage systems, practically for all field conditions. Several of these solutions in this chapter have been published previously but this book makes the list complete. The chapter 4 gives mathematical models on fluctuating water tables in response to crop drainage requirements and the efficiency of the designed drainage system. Determination of drain spacing is calculated using fluctuating water table models by predicting water table depth in response to rainfall or recharge or irrigation.

I think this book makes a nice addition to the university libraries and will benefit faculty and students. This examines drainage design formulas and efficiency of surface and subsurface drainage systems. This book gives several working examples and references to help student lean on how to design drainage systems. Any reader interested in drainage and design of drainage systems will find this book very interesting and enlightening for many years to come.

RAMESH KANWAR
Professor of Agricultural and Biosystems Engineering
Iowa State University, Ames, Iowa 50011
U.S.A.

PREFACE

This book is intended to serve as a text or a reference book for engineering students studying at a senior level in their college or university education. The main purpose of this book is to present theories, principles, and approaches to the design of drainage systems. The history of agricultural drainage design dates back to the Greek historian Herodotus in 400 B.C. with reference to the Nile Valley in ancient Egypt. Many Roman authors (Cato, Virgil and others) also have given reference to agricultural and urban drainage design systems using channels in the early eastern nations. Depending on the climatic, hydrology, landscape, crops, soil types and management factors, different types of drainage design methods have been used for different agricultural fields during the history of mankind. This book does an outstanding job in compiling all drainage design methods and associated drain design formulas. Several boundary value problems have been developed for different field conditions and the needed solutions were derived using mathematical models so that practicing engineers can use these mathematically derived solutions to design surface and subsurface drain systems.

This book addresses special drainage needs for the irrigated agricultural fields. Irrigated agriculture increases crop yields and economic health of farmers but it also could raise shallow ground-water tables and salinity levels because of excessive irrigation. Therefore, properly designed irrigation systems for agricultural crops must have adequately designed drainage systems. Next question is: what should a successful drainage system have to promote its long-term effectiveness or sustainability in the agro-ecosystem? Depending on the climate, hydrogeology, soil characteristics, topography, farming practices, the drainage effluents and their effects on soil and water quality need to be managed. The disposal and management of drainage effluents from agricultural fields need special attention to minimize its offsite effects on the environment. This book will serve as a useful reference book to answer some of these questions.

This book has five different chapters, chapter 3 to be the largest and most comprehensive in material and thoughtfulness. The chapter 1 is an introductory chapter giving definitions of various types of drainage and needs of drainage for agricultural lands. This chapter sets the foundation for writing rest of the chapters of this book. The chapter 2 is written on "saturated hydraulic conductivity". This chapter begins with a definition of conductivity and then gives extensive and clear details on laboratory and in-situ field methods to measure saturated hydraulic conductivity. This chapter also describes theories behind these methods used for the determination of hydraulic conductivity. For each method several working examples are given for the benefits of the students and to bring a better understanding on the use of these methods to determine saturated hydraulic conductivity.

The chapter 3 gives steady-state unsteady-state solutions for boundary value problems for field collector and main drains using different types of flow domain systems such as cylindrical, elliptical and spherical. Drainage theories by Dupuit-Forchheimer and Hooghoudt (1937) have been used to develop drainage models for sloppy lands and layered soil profiles. Also, theories developed by Khirkham and Grover have been used to determine drain spacing for soils with varying hydraulic conductivity. In addition this chapter provides solutions for steady-state and unsteady-state drainage systems under fluctuating water table conditions. In reality, this chapter provides methods for the design of drainage systems, practically for all field conditions. Several of these solutions in this chapter have been published previously but this book makes the list complete. The chapter 4 gives mathematical models on fluctuating water tables in response to crop drainage requirements and the efficiency of the designed drainage system. Determination of drain spacing is calculated using fluctuating water table models by predicting water table depth in response to rainfall or recharge or irrigation.

I think this book makes a nice addition to the university libraries and will benefit faculty and students. This examines drainage design formulas and efficiency of surface and subsurface drainage systems. This book gives several working examples and references to help students learn on how to design drainage systems. Any reader interested in drainage and design of drainage systems will find this book very interesting and enlightening for many years to come.

CONTENTS

Chapter 1

INTRODUCTION

Irrigation is the artificial application of water to the soil to supplement the water available from rainfall and contribution to soil moisture from ground water for the purpose of beneficial crop production.

Irrigation water is a key input for improved agricultural production. It is an age-old practice. Irrigation has played a major role in the development of old civilisation. Ancient civilisation developed along Nile, Tigris, Euphrates, Indus and Yellow rivers. Gravity irrigation began along the Nile about 6000 B. C. The common methods of irrigation from early times have been surface or gravity and sprinkler irrigation.

Irrigation on one side has several advantages, probably without irrigation whole civilisation may vanish. Government of India has given due emphasis to increase water resources of the country. In 1940s, India barely had two big storage reservoirs with storage capacity exceeding one billion cubic meters. Since then, 28 large storage had been built. Plan-wise irrigation potential created and utilised is given in **Table 1.1.** The water resources in the country are not distributed uniformly. For example 29 per cent of the water resources are available in the Brahmputra basin, which constitutes only 6 per cent of the country's area. Southern and Western regions have much lesser water resources than the national average. In some areas availability is as low as one fourth of the national average. In the year with abnormal low rainfall the water resources scenario may become quite serious. Government of India has given due emphasis on creating additional water resources to bring more and more area under irrigation. **Table 1.2** shows the irrigated area by different sources in different years. Canal irrigated area has increased significantly since independence and still covers the maximum area under irrigation.

Food grain production has increased significantly with irrigation but to the

Table 1.1 Plan-wise irrigation potential created and utilised in India (cumulative)

Plan & period	Potential created, million hectare					Potential utilised, million hectare					Percent utilisation
	Major & Medium	Minor			Total	Major & medium	Minor			Total	
		Surface water	Ground water	Total			Surface water	Ground water	Total		
Pre-plan Upto 1951	9.70	6.40	6.50	12.90	22.60	9.70	6.40	6.50	12.90	22.60	100.00
I plan 1951-56	12.30	6.43	7.63	14.06	26.36	10.98	6.43	7.63	14.06	25.04	95.00
II plan 1956-61	14.33	6.45	8.23	14.73	29.06	13.05	6.54	8.28	14.82	27.87	96.00
III plan 1961-66	16.57	6.48	10.52	17.00	33.57	15.18	6.50	10.52	17.02	32.20	96.00
Annual plan 1966-69	18.10	6.50	12.51	19.00	37.10	16.75	6.51	12.51	19.02	35.77	96.00
IV plan 1969-74	20.70	7.00	16.44	23.44	44.14	18.69	6.96	16.44	23.40	42.09	95.00
V plan 1974-78	24.72	7.50	19.80	27.30	52.02	21.16	7.50	19.80	27.30	48.46	93.00
Annual plan 1978-80	26.61	8.00	22.00	30.00	56.61	22.64	8.00	22.00	30.00	52.64	93.00
VI plan 1980-85	27.70	9.70	27.82	37.52	65.22	23.57	9.01	26.24	35.25	58.82	90.00
VII plan 1985-90	29.92	10.90	35.62	46.61	76.63	25.47	9.97	33.15	43.12	68.59	90.00
Annual plan 1990-92	30.74	11.46	38.89	50.35	81.09	26.32	10.29	36.25	46.54	72.86	90.00
Annual plan 1992-93	31.13	11.62	40.26	51.88	83.01	26.62	10.40	37.49	47.89	74.51	90.00
VIII plan 1992-97	32.70	12.20	45.90	58.10	90.80	28.20	10.80	42.10	52.90	81.10	90.00

Table 1.2 Area (thousand ha) irrigated by different sources in India

Source	1950-51	1960-61	1970-71	1980-81	1984-85	1992-93*
Canals	8295	10370	12838	15292	15861	17084
	(39.8)	(42.1)	(41.3)	(39.5)	(37.9)	(34.0)
Tanks	3613	4561	4112	3198	3330	3243
	(17.3)	(18.5)	(13.2)	(8.2)	(7.9)	(6.4)
Tubewells	NA	135	4461	9527	11265	15824
		(0.6)	(14.3)	(24.6)	(27.9)	(31.8)
Other wells	978	7155	7426	8207	8723	10714
	(28.7)	(29.0)	(23.9)	(21.1)	(20.9)	(21.4)
Other sources	2967	2440	2266	2585	2600	3236
	(14.2)	(9.8)	(7.3)	(6.6)	(6.2)	(6.4)
Total	20853	24661	31103	38806	41779	50101
(Net irrigated area)	(100)	(100)	(100)	(100)	(100)	(100)

Figure in bracket are the percent of total irrigated area
* provisional

contrary, various demerits are also associated with irrigation. Excessive and injudicious use of irrigation water has created waterlogging and salinisation all over the world. The growth of a plant depends on the extent and vigour of its root systems. When soil remains saturated for longer period the structure breaks down and size of pore space becomes smaller and uncongenial for root growth. Root will not thrive if the pore space is full of water nor can bacteriological and other process, on which well being of plant depend function properly. Where the land is waterlogged most of the plants roots tend to grow near the land surface and their extent is restricted by the small amount of soil volume available for nutrient extraction.

When irrigation is introduced on a large scale into un-irrigated semi-arid and arid areas without adequate drainage measures, the water table rises and causes waterlogging problem. The semi-arid and arid areas in general have salts in their soil profiles and ground water. The rising water table brings these salts upward and waterlogging results in secondary salinisation, when water table reaches the root zone water starts evaporating leaving these salts behind the ground surface. In humid areas where annual rainfall is much higher than the average annual evaporation, the rate of downward movement of these salts is higher than the upward movement and no salinity develops due to leaching.

1.1 Definition of Waterlogging

Waterlogging is defined as the retention of areas under water for a considerable period of time, causing severe damage or complete failure of crop. An area is said to be waterlogged when the water table rises to an extent that the soil pores in the root zone of the crop becomes saturated resulting in restriction of normal circulation of air, decline in the level of oxygen and increase in the level of carbon dioxide. There is a lack of a common definition of waterlogging. A number of definitions of waterlogging are being used by different organisation.

(1) The National Commission on Agriculture defines waterlogging as "An area is said to be waterlogged when the water table rises to an extent that the soil pores in the root zone of a crop becomes saturated resulting in restriction of the normal circulation of air, decline in the level of oxygen and increase in the level of carbon-di-oxide".

(2) Central Board of Irrigation and Power (CBIP) defines waterlogging as "An area is said to be waterlogged when the water table rises to an extent that the soil pores in the root zone of a crop becomes saturated resulting in restriction of the normal circulation of air, decline in the level of oxygen and increase in the level of carbon-di-oxide. The water table, which is considered harmful, will depend on the type of crop, type of soil and the quantity of water. The actual depth of water table when it starts affecting the yield of the crops adversely may vary over a wide range from zero for rice to about 1.5 m for other crops". The norms of waterlogging suggested by CBIP are given below.

Nomenclature	Depth of water below ground surface
Waterlogged	< 2 m
Potential area for waterlogging	2-3 m
Safe	>3 m

(3) Central Ground Water Board (CGWB) defines waterlogging as that hydrological condition when ground water table reaches the shallow critical and semi-critical conditions and soil profile becomes fully or partially saturated with capillary ground water. For different soil groups and different crops/trees, the critical and semi-critical water level is different because of different limits of capillary rise of ground water as given below.

Table 1.3 Capillary rise in agriculture soils

S.N.	Soil type	Capillary rise (in cm)
1.	Sandy loam	100-150
2.	Light loam	150-200
3.	Heavy loam	300-500
4.	Clay	400-500

CGWB classified waterlogging in four categories are given below.

(a) Ponded/flooding conditions

In this condition the water is accumulated on the surface temporarily or for a longer duration and in most cases ground water level reaches above the ground. Drainage of such soils is not feasible because of reverse/smooth gradient.

(b) Marshy lands

In this condition, for most of the year ground water level is above the ground surface and for part of the year less than 0.5 m below ground level, such a condition can be tolerated by many crops and thus need to be maintained for such crops. In the presence of salts or with poor quality water, deeper root zones need be clear of water and salts for proper plant growth and thus such condition is also required to be maintained.

(c) Critically waterlogged

These are the areas where the ground water reaches the surface with capillary rise. This height varies for different soils but may be taken as 1.5 m to 3.0 m below the ground level.

(d) Semi-critically waterlogged

In such areas the capillary rise of ground water reaches the root zone for most of the year, not always to the surface. The semi-critical waterlogged area may be taken as 3.0 m to 6.0 m below ground level.

1.2 Norms of Waterlogging in Different States

Different states are using different norms of waterlogging as given in **Table 1.4**.

Table 1.4 Waterlogging norms used by different states in India

States	Water table depth below ground level
Haryana	
a) Critically waterlogged	0.0 – 1.5 m
b) Moderately waterlogged	1.5 – 3.0 m
Himachal Pradesh	0.0 – 2.0 m
Karnataka (Tungbhadra)	0.0 – 2.0 m
Maharashtra	
a) Fully waterlogged	water at surface
b) Waterlogged	0.0 – 1.2 m
Punjab	
a) Very critical	0.0 – 1.5 m
b) Critical	0.0 – 2.0 m
Uttar Pradesh (Sharda Sahayak)	
a) Worst zone	< 1.0 m
b) Bad zone	1.0 – 2.0 m
c) Alarming zone	2.0 – 3.0 m
d) Safe zone	> 3.0 m

It may be seen that there is no uniform norm/definition of waterlogging. In the lack of a common definition various estimates of waterlogging are different and creating confusion.

1.3 Extent of Waterlogging

Extent of waterlogging in some of the large irrigated areas is shown in **Table 1.5**. China and India, most populous countries have got the largest area under waterlogging.

Table 1.5 Waterlogging and salinity in countries with large irrigated areas

Country	Irrigated area (million ha)	Total waterlogged and salt affected area (million ha)	Extent of malady as per cent of irrigated area
China	47.8	11.0	23.0
USA	18.1	5.20	29.0
Pakistan	15.4	3.20	21.0
Iran	5.9	1.72	30.0
Mexico	6.0	0.6	12.0
India	47.8	6.0	12.0
Egypt	2.7	0.88	32.7

There is a great variation among the reported waterlogged areas from different sources in India **(Table 1.6)**. There is a lack of methodically monitored information on the type, severity and extent of waterlogging in the country.

National Irrigation Commission (1972) estimated that 4.75 million hectare of land were affected by waterlogging in the States of Panjab, Haryana, Uttar Pradesh, Bihar, West Bengal, Maharashtra, Rajasthan, Madhya Pradesh and Karnataka. National Commission on Agriculture indicated that 3.4 million hectare of land was affected by flooding and about 2.6 million hectare suffered with high water table conditions. **Johari et al. (1990)** reported that by the year 1990 waterlogging affected about 6 million hectare of land. Extent of waterlogging in different states of India is given in **Table 1.7**.

Table 1.6 Various estimates of waterlogged area in the country.

Source	Waterlogged area (million hectare)
Irrigation Commission (1972)	4.75
National Commission on Agriculture (1976)	6.00
Ministry of Agriculture (1984-85)	8.53
Working Group of Ministry of Water Resources (1991) only in irrigated command	2.46
Singh (1994)	4.50
Abrol (1994)	16.00
Singh and Bandyopadhyay (1996)	3.95
Sehgal (1996)	5.20

Table 1.7 Extent of waterlogged area as estimated by Ministry of Water Resources Working Group

State	Affected area
Andhar Pradesh	266.40
Assam	NR
Bihar	619.70
Haryana	249.00
Jammu and Kashmir	150.20
Himachal Pradesh	0.20
Karnatka	24.54
Kerala	11.61
Madhya Pradesh	73.12
Maharashtra	15.35
Orissa	196.26
Punjab	200.00
Rajasthan	175.50
Tamil Nadu	16.59
Uttar Pradesh	430.00
West Bengal	NR
Union Territories	NR

Bihar and Uttar Pradesh is having the largest area under waterlogging in the country.

1.4 Causes of Waterlogging

The rapid development of irrigation during the past 100 years, to increase food grain production to feed increasing population has brought in acute problems of waterlogging and salinity in South Asia and Middle East. All over the world, wherever the large scale irrigation has been introduced without giving due consideration to drainage or any built-in anti-waterlogging measures, extensive waterlogging and salinisation has occurred. Following are the points identified for waterlogging.

1. High rainfall.
2. Seepage from earthen dams, canals, irrigation channels etc.
3. Percolation of irrigation water from the field.
4. Partial or full obstruction of natural drainage of rain water by other development like roads, railways and so on.
5. Introduction of high water demanding crops in arid and semi-arid areas.
6. Excessive and non-judicious use of irrigation water.

1.5 Salt Affected Soils

When water table rises due to excessive seepage or deep percolation or due to any other reason, it dissolves salts present in the deeper soil profiles and brings near the soil surface in due course of time, where water moves through the capillary and gets evaporated due to solar radiation leaving the salts on the ground surface. Slowly soil becomes saline and unfit for crop productions.

In India salt affected soils have been known to occur since time immemorial, but recorded evidence is available since the middle of nineteenth century. The salinity problem received attention in 1886 when the Government of India deputed Engineer J.W. Wilson to the USA for holding discussion with E.W. Hilgard, a pioneer salinity researcher. J.W. Leather, then the Imperial Chemist, was the first scientific worker in India who initiated research on possible causes of salinisation. The scientific investigations continued to be made in the first half of twentieth century, generating information on the relative salt tolerance of crop plants, methods of *usar* reclamation and drainage of waterlogged saline lands. There exists a strong relationship between the electrical conductivity (EC) of soil extract and the soil's salt concentration, hence the salt content of the soil is commonly expressed by EC measured at reference temperature of 25 °C. The EC is expressed in dS/m (decisiemens per meter) which is equivalent to older unit millimhos per centimeter (mmho/cm).

U.S. Salinity Laboratory (**Richards, 1954**) has classified salt affected soils as under.

I. Non-Saline Alkali/Sodic soils

These soils contain excess of salts capable of alkaline hydrolysis such as sodium carbonate, sodium bicarbonate and sodium silicate; and sufficient exchangeable sodium to impart poor physical conditions to soils and thus affecting growth of most plants. These soils have an EC<4 dS/m, at 25°C and an ESP>15. The pH is usually higher than 8.5; a pH of 10 or even more is no exception. The pH of the calcareous alkali soils is highly related to their ESP. Presence of calcium carbonate concretions at about one meter depth causes physical impedance for root proliferation. The soil structure of nonsaline sodic soils can often be regarded as poor.

II. Saline soils

These soils with white salt encrustation on the surface have predominantly chlorides and sulphates of Na^+, Ca^{++} and Mg^{++}. HCO_3^- is present in small quantity and NO_3^- is rarely found. Na^+ as a rule comprises less than 50% of the soluble cations. The calcium carbonate and gypsum may be present. The soils with neutral soluble salts have saturation paste pH<8.5. Electrical conductivity of saturation extract, ECe is generally more than 4 dS/m at 25° C and ESP<15. Such soils invariably have sodium absorption ratio (SAR) of the soil solution>15. In saline soils the excess of neutral salts restrict the normal growth of the plants. Due to presence of excess salts, saline soils usually remain flocculated and their hydraulic conductivity is similar to those of non-saline soils. The main causes of poor plant growth in saline soils are high osmotic pressure of soils solution, causing low physiological availability of water to the plant, direct toxic effects of individual ions and complex interaction between sodium, calcium and magnesium, leading to disturbed equilibrium of these ions in the plant's ability to absorb water and nutrients in required amounts.

III. Saline sodic soils

Saline sodic soils have an ECe >4 dS/m at 25° C and an ESP>15. The pH is seldom higher than 8.5. Often saline sodic soils have a pH value near to neutral. The sodium ions in solution are present as neutral salts as NaCl and Na_2SO_4. If the pH value is above 8.5, the ions HCO_3^- and CO_3^{--} are present in soil solution. Such saline sodic soils tend to be more problematic to reclaim.

1.6 Extent of Salt Affected Soils

Extensive salinisation has occurred all over the world where large-scale irrigation development has taken place without giving due consideration to the drainage of the area. **Table 1.8** shows the country-wise irrigated area and per centage of area which has become saline all over the world.

Table 8. Country-wise per cent irrigated area, which has become saline

Country	Area irrigated (thousand ha)	Per cent area affected by salinity	Country	Area irrigated (thousand ha)	Percent area affected by salinity
Algeria	296	10-15	Iraq	1750	50
Egypt	2474	30-40	Israel	220	13
Senegal	175	10-15	Jordan	36	16
Sudan	1700	<20	Pakistan	15320	<40
USA	19831	20-25	Sri Lanka	550	13
Columbia	322	20	Syria	617	30-35
Surinam	45	12	Greece	1030	7
China	45420	15	Portugal	632	10-15
Cyprus	94	25	Spain	3145	10-15
India	39700	27	Australia	1628	15-20
Iran	5730	<30			

About 10 to 50% of the irrigated area has become saline in different parts of the world.

The area under salt affected soils was estimated to be 7 million ha by **Abrol and Bhumbla (1971).** However, during the last decade several agencies have given divergent estimates, e.g. National Commission on Agriculture, 7.16 million ha; National Remote Sensing Agency, 3.9 million ha; National Wasteland Development Board, 1.5 million ha, National Bureau of Soil Survey and Land Use Planning, 6.2 million ha and **Singh (1992)**, 8.6 million ha. Modified and critically evaluated data by **Central Soil Salinity Research Institute, Karnal (1996)** are given in **Table 1.9**.

Table 1.9 Extent and distribution of salt affected soils in India (thousand ha)

State	Waterlogged area			Salt affected area			
	Canal	Non-Command	Total classified	Canal command	Outside canal	Coastal	Total
Andhra Pradesh	266.4	72.6	339.2	1239.4	390.6	283.3	813.3
Bihar	362.6	NA	362.6	224.0	176.0	NIL	400.0
Gujarat	172.6	311.4	484.0	540.0	327.1	302.3	455.0
Haryana	229.8	45.4	275.2	455.0	NA	NIL	455.0
Karnataka	36.0	NA	36.0	51.4	266.6	86.0	404.0
Kerala	11.6	NA	11.6	NA	NA	26.0	26.0
Madhya Pradesh	57.0	NA	57.0	220.0	22.0	NIL	242.0
Maharashtra & Goa	6.0	105.0	111.0	446.0	NA	88.0	534.0
Orissa	196.3	NA	196.3	NA	NA	400.0	400.0
Punjab	198.6	NA	198.6	392.6	126.9	NIL	519.5
Rajasthan	179.5	168.8	348.3	138.3	983.8	NIL	1122.0
Tamil Nadu	18.0	109.9	127.9	256.5	NA	83.5	340.0
Uttar Pradesh	455.0	1525.6	1980.0	606.0	689.0	NIL	1295.0
West Bengal	NA	NA	NA	NIL	NA	800.0	800.0
Total	2189.4	2338.1	4527.5	3469.1	3027.0	2069.1	8565.2

1.7 Definition of Drainage

Agricultural drainage as indicated above is an effective method to maintain sustainable agricultural system. The definition of land drainage given in the constitution of International Commission on Irrigation and Drainage/**ICID (1979)** reads as follow.

"**Land drainage is the removal of excess surface and subsurface water from the land to enhance crop growth, including the removal of soluble salts from the soil**"

The main purpose of drainage is to provide a favourable environment that is conducive for the maximum plant growth. The object of providing drainage is to increase production and sustain yields over long period of time. Many times the objectives of the drainage may be to bring additional area under cultivation or/ and to advance sowing time of next crops.

1.8 The Need for Land Drainage

The world population, at present, is estimated as about 500 million. The average annual growth rate in the world population is approximately 2.6%. Half of the world population lives in the developing countries where population growth rate is higher and demand for food and fibre is growing rapidly. Large scale irrigation has brought waterlogging and salinity problems all over the world and a sizeable area is going out of cultivation and productivity in irrigated area is declining constantly. Productive agricultural lands are going out of cultivation due to rapid urbanisation in developing countries. More than 50 per cent of the canal irrigated areas are affected by waterlogging and salinity problems in varying degree due to improper distribution, application and management of irrigation water. Share of fresh quality water to agriculture sector is also shrinking. Land drainage or the combination of irrigation and land drainage is one of the most important input factors to maintain or to improve yield per unit of farmed land. **Figure 1.1** shows the impact of irrigation water management and the control of water table.

Land drainage seems to be a tool to increase agricultural production to cater to the demand of growing population by:

- preventing a decrease in the productivity of arable land due to rising

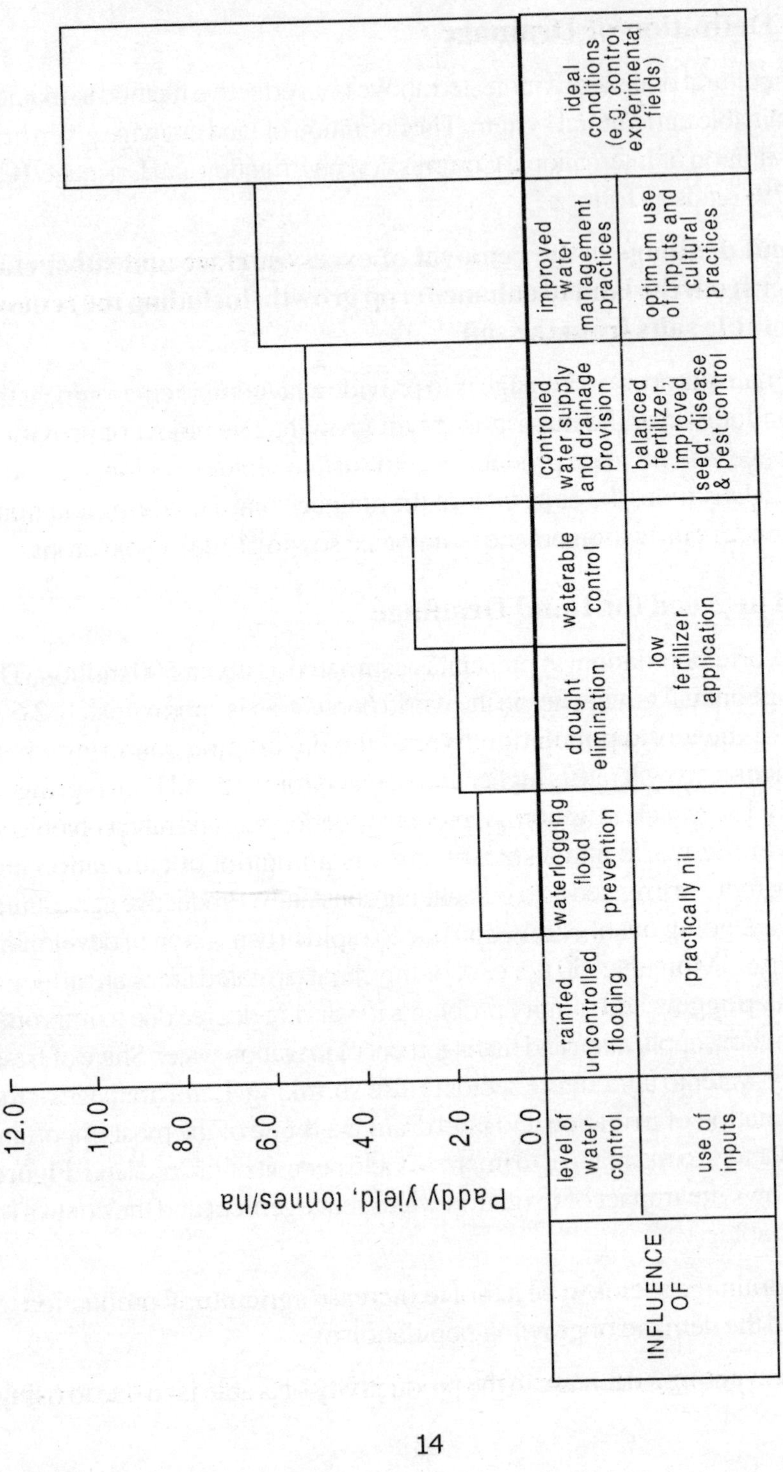

Fig. 1.1. Effect of improved water management and inputs on paddy yield.

water tables and salt build-up in the root zone,

- bringing additional area under cultivation by lowering the water table and removing the standing water from the area,
- reclaiming salt affected soils (sodic or saline) and putting under agricultural use, and
- utilising applied nutrients and other inputs efficiently.

1.9 History of Drainage in India

Waterlogging and salinity was first noticed in the upper region of Rechana Doab, a few years after the opening of lower Chenab canal in 1892, whereas the problem of drainage and salinity were observed in the Meenak area of Karnal district in 1855. The Punjab Government made an attempt for the control of salinity in the form of a survey carried out in lower Chenab colony in 1908. In early 1925 the Punjab Government constituted a committee to study and report on the extent and cause of waterlogging. Based on this a small farm was set up at Chankanwali for the field experiments for reclamation of waterlogged areas. In 1928, the waterlogging committee was abolished and a waterlogging board was constituted. Since independence the land reclamation, irrigation and power research institute at Amritsar had taken up this work.

Inglis and Gokhale (1928) conducted first drainage studies at Baramati experimental far with deeper drains to release the confirmed subsurface water into the drain. It was reported that these drains worked effectively in lowering the water table and reducing the salt contents. In 1930 the Manjari scheme was initiated near Pune which consisted of a network of the drains installed in agricultural lands. It was found that water table depths and salinity levels were reduced and an economic level of crop production was obtained.

As reported by Narmada Project preparation cell in Saurashtra, a pilot polder project was constructed in 1959-61, with the cooperation of The Netherlands for reclamation of saline lands in coastal plains. Drains were installed at 0.7 m and 0.8 m depths at spacing of 25, 50 and 92 m. The results obtained were poor. For similar works better results were obtained at Kharland research project for polder in Maharashtra.

At Punjab Agricultural University, Ludhiana in 1972, some works were carried out in developing drainage criteria in 1966. At Siruguppa (1969-70) studies were carried out on black soils by laying 10 cm tile drains at 12, 24 and 36 m spacing and 0.8, 1.0, 1.1 and 1.4 m depth at 0.3 per cent gradient. They tried open drains, covered with leaves, 10 cm diameter pipes, 15 cm diameter clay holed tiles, 10 cm diameter cement holed pipes and 15 cm diameter clay holed pipes in their treatment for finding best type of subsurface drainage in the black soil.

At Jawaharlal Nehru Krishi Vishwa Vidyalaya at Indore in 1970-71 an experiment was conducted to study the effect of surface drainage, subsurface drainage and a combination system on crops. The study was made in a combination of 60, 75 and 90 cm drain depths and spacing of 6, 18, and 36 m, by laying the drains of 10 cm diameter made of semi glazed stone and cement asbestos pipes by laying the drains in diagonal system.

Experiments were carried out at Central Soil Salinity Research Institute, Karnal in Alkali soils laying drains at 1.5 m depth and spacings 10, 20 and 30 m. The tile drains and open drains were laid for the study. **Chauhan and Sewa Ram (1972)** studied combined effects of surface and subsurface drains at 15, 25 and 35 m spacing and one meter depth for two rates of replenishment. The studies were carried out in humid lands of Pantnagar Tarai.

Central Soil Salinity Research Institute carried out studies at National Dairy Research Institute Farm at Bidaj in 1976 to carry out the reclamation of saline soils in heavy textured clay loam soils where subsurface drains of 15-25 m spacing were laid at 1.5 m depth in Parbhani clay soil in vertisols of Marathwada region. Experiments were also conducted in 1983-84 at Central Soil Salinity Research Institute, Karnal. Rao and Kamra (1983) studied at CSSRI in an area at Sampla with pumped outlet by laying subsurface drains at 1.75 m depth and 25, 50 and 75 m spacing using cement concrete tiles of 10 cm diameter and 30 cm length with an objective to develop suitable criteria for design of subsurface drainage for controlling waterlogging and salinity.

Works were also carried out at Gohana by CSSRI, Karanal in 1984. The drains were laid at 1.5 m to 1.6 m depths and 50 m spacing with a gravity outlet. Works were also carried out by CSSRI, Karnal in collaboration with Haryana Land Reclamation Development Corporation and studying susbsurface drainage in an area of about 8 ha and 16 ha by laying drains at spacings 50, 76 and 84 m at a depth of 1.75 m.

Studies were carried out at Haryana Agricultural University and other institutions. It was observed that vertical drainage tube wells of diameter 10 cm to 20 cm and depth of 25 to 40 m, discharge 40 to 80 m^3/hour can maintain a ground water level at 2 to 4 m below ground surface for 200 to 300 days pumping in a year with proper spacing. Subsurface drains laid at 25, 50, 67 and 75 m spacing and depth 1 to 2 m, in 10 to 60 ha has provided high crop productivity.

In Gujarat large areas in Mahi Command and Sardar Sarovar are waterlogged. Studies at Water and Land Management Institute, Gujarat alongwith Central Soil Salinity Research Institute, Karnal were done on horizontal drainage of laying drains at 1.2 to 1.5 m depth and 50 m spacing for studying proper filters. Out of different filters like geo-textile plastinet and coir, gravel was found to be the best.

In acidic soils of Kerala studies of horizontal drainage were done by laying drains at 15 m and 30 m spacing by pipe made of baked clay. The subsurface drains were successful in removing toxic acidic substances improving productivity.

The Siruguppa studies showed that tile drains gave better results than other types of drains consisting of boulder and other materials. The Indore studies showed that the drains laid at 6.5 m spacing and 60 cm depth gave reasonable maximum yield as compared to other treatments. Studies at Pantnagar provided useful hydrological data for the area and indicated that 25 m spacing appeared to be appropriate for humid lands in Tarai. The studies at Ludhiana showed an effective lowering of water table by about 90 cm and that subsurface drainage appeared to provide better yield than control. Studies carried out at Central Soil Salinity Research Institute, Karnal lowering ground water table and reclamation of saline soils, such that good crops could be grown in the land where nothing was grown earlier.

1.10 Effects of Poor Drainage on Soil and Plants

1. Waterlogging restrict the aeration and exchange of gases between roots and atmosphere. The availability of oxygen is limited due to small soluble amount in water and extremely low rate of diffusion through waterlogged soils.

2. In absence of dissolved oxygen anaerobic decomposition of organic matter starts which results in production of methane, methyl compounds and complex aldehydes. Mineral substances in the soil are altered from oxydised state to reduced state. Toxic concentrations of ferrous and sulphide ion may develop within a few days after submergence.

3. Nitrogen tends to remain locked up in the organic residues due to slowing down rate of decomposition.

4. Transpiration rate is declined and plants experience difficulty in taking up moisture under waterlogging conditions. This is due to decreased concentration of oxygen and increased concentration of carbon-di-oxide. These conditions persist due to the decay of roots and the lack of formation of new roots.

5. Under waterlogging condition nutrient supply is disturbed and plant may show yellowing, reddening, or a scorched or stippled appearance of leaves.

6. The ponding of water during hot season may cause scalding and planted grasses or legumes may die.

7. Mosquitoes, snails and other insects may create various health hazards in waterlogged areas.

8. Excessive soil moisture is often the cause of soil compaction by animals and machines.

9. High water table may result in slow warming of soil and germination may be delayed. The seed may rot before germination.

10. Plant diseases are more active under waterlogged conditions. Fungus growth is particularly prevalent.

11. **Accumulation of salts can convert most fertile lands into barren desert in irrigated areas. This is one of the biggest problems in irrigated agriculture.**
12. Water table build-up in irrigation areas may change the local environment.

1.11 Advantages of Land Drainage

Solutions to the drainage problems have indicated several benefits in different projects. These benefits are given below.

1. Better soil aeration results from good drainage. This promotes extensive root development and gives conducive environment for beneficial soil micro-organisms and earth warms. This may further reduce the root diseases.
2. Better soil moisture with good drainage permits more efficient operation of tillage, planting and harvesting equipment.
3. Better soil structures can be developed and maintained with good drainage since there is less chance of destroying soil tilth due to compaction when working soil is too wet.
4. Soil warms up more quickly when free water is removed by a drainage system. This results in better germination and an increased rate of plant growth.
5. An increased supply of nitrogen can be obtained from the soil when drainage lowers the water table in the root zone; de-nitrification often occurs in soils with poor drainage.
6. Larger growing season can be achieved with good drainage due to earlier possible planting dates.
7. Certain toxic substance and disease organisms are removed from the soil due to better drainage and soil aeration. In wet soil, roots can be injured by toxic substances produced in the reduction of iron and manganese salts and the reduction of nitrates to nitrites.
8. Winds are less liable to uproot plants growing in soils that have properly been drained, since root systems are deeper, and soil strength is greater.

9. Soil erosion and sediment loss can be reduced by subsurface drainage, since drained soils have a greater capacity to absorb rainfall and the soil filters out suspended sediment.

10. Good drainage saves fuels that would be used in working around wet areas in fields that are not properly drained. Since drained land is easier to work there is less need for dual wheels or four wheel drive tractors besides increasing the life of machining and equipment.

11. A good variety of crops can be grown successfully over a well drained farm land.

12. Weed problem is minimised and its control is easier with good drainage since shallow rooted weed and undesirable grasses often thrive in wet soils crowding about the planted crops.

13. Well drained grazing land supports more livestock, with less compaction damage to vegetation and soil from animal traffic.

14. Good drainage reduces diseases that thrive on wet land. These include footrot, liver fluke that infect livestock, and diseases carried by mosquitoes to both livestock and human beings. The plant diseases such as root rot, stem rot etc. are also reasonably controlled.

15. Drainage is essential for salinity-alkalinity control in arid/semi-arid regions.

1.12 Selection of a Drainage System

Drainage can be either natural or artificial. Most of the lands have some natural surface and subsurface drainage. In case the natural drainage of the area is inadequate to remove excessive water reaching the land either through rainfall i.e. natural means or by seepage from reservoirs, canals and excessive irrigation i.e. artificial means, manmade or artificial drainage is required. Actually every area has natural drainage system sufficient to cope with normal climatalogical responses. Man made activities are checking natural drainage system or delivering excess water in the area than it can drain naturally or both. Artificial drainage thus fills the gap between that provided by nature and the present need.

The drainage planner for a drainage system in an irrigated area must be constantly aware of land and water resources as well as agricultural aspects. Environmental aspects should also be given due consideration in planning drainage system. Solution of drainage system can be made after careful examination of various alternating methods.

Important drainage methods are given below:

1. Surface drainage including land grading and smoothing.
2. Subsurface drainage
   a) Horizontal or pipe drainage.
   b) Vertical or tube well drainage, and
3. Bio-drainage.
4. Preventive measures like,
   a) Lining of water distribution systems.
   b) Improved on farm water management practices.
   c) Maintenance of existing drainage system.
   d) Introduction of advanced irrigation methods like drip.
   e) Avoiding introduction of high water demanding crops in the area.

1. Surface drainage

ASAE (1979) describes surface drainage as "the removal of excess water from the soil surface in time to prevent damage to crops and to keep water from ponding on the soil surface, or, in surface drains that are crossed by farm equipment, without causing soil erosion". The water may be from excess precipitation, water applied in irrigation, losses from conveyance channels and storage systems or water which has seeped from ground water at higher elevation. Flat or level land, land having too small slope or land having impermeable sub-soils with shallow top soil often requires surface drainage because subsurface drainage may not be practical or economical. Wherever mechanised farming operations are done, surface drainage system should meet its requirement. Surface drainage or also required sometimes in conjunction with subsurface drainage for drainage sensitive crops. Open drains are advantageous for removing large volumes of either surface or subsurface drainage water including storm water from land where that water table is near the ground surface. Open drains have got same disadvantages like,

a) Take land out of production.

b) Cause soil erosion.

c) Require frequent cleaning.

d) Harbour and distribute obnoxious weeds.

Surface drainage broadly comprises five categories.

I. **On farm field drainage system:** It consists of graded channels that collect excess water from field.

II. **Intermediate drains (Collector or carrier drains)**: There are link drains between various field drains and sub-main drains or main drains.

III. **Main drains**: Main/sub-main drains are principally excavated or natural drains, collecting water from link drains or directly from field drains.

IV. **Seepage drains**: Seepage drains are located along earth/lined canals and interrupt surface/subsurface flow. These drains help in management of excess discharge from filled reaches of canals to the adjacent fields.

V. **Land grading and smoothing**: Most natural soil surfaces including many cultivated fields are more or less irregular consisting of randomly distributed elevated areas and depressions. It results in pondage of water, inefficient irrigation, interrupted surface drainage, difficulties in mechanized field operation and consequently non-uniformity of crop yield and poor quality grain. In many cases these areas, may be relieved of these problems by reforming the land surface to provide gentle slope for resultant regulated water flow and adequate drainage. Land grading and smoothing is probably the simplest yet the most important type of surface drainage.

2. Subsurface drainage

Subsurface drainage is the "removal of excess soil water in time to prevent damage to crops because of a high groundwater table" and is accomplished by controlling water table positions.

a) Horizontal subsurface drainage

This is also called as pipe drainage and consists of pipes made of clay, concrete, and plastic or of any suitable perforated conduit generally laid at a depth of 1 to 3 meter below ground surface. Excess water enters the pipelines through the perforations or through the joints in concrete or clay tiles and flows towards main drains through collector.

Pipe drains are used mainly for

- Intercepting the lateral flow of ground water coming from a known source of seepage up- slope, to keep it from reaching the area to be protected,
- Lower the ground water table and to remove the excessive soil moisture from the plant root zones over the drainage area.
- To leach down the excessive salts present in root zone profile.

Sometimes mole drainage is also practised under suitable soil conditions. Mole drains are the pipeless cylindrical channels made artificially in the ground through a mole plough.

b) Vertical drainage

This is also called as tube well drainage and can be described as "control of an existing potential high groundwater table or artesian ground water conditions". Under certain condition pumping of ground water provides an effective land drainage method for lowering water table and reducing salt hazards in irrigated areas. The vertical drainage consists of a battery of tube wells spread over the area to be drained. In many cases, if water quality is good, pumped water can be recycled for irrigation.

3. Bio-drainage

Use of quick growing and high water demanding plants (exotic trees) have been utilised to improve marshy and waterlogged areas. These plants extract water at a faster rate from the soil. This system of drainage requires thorough investigations on plant species to suit specific areas particularly in saline soils. Trees such as eucalyptus have been shown to trap up to 25-30 mm of rainfall and evaporate about 25%, thereby preventing its movement into the soil profile. However, it is unlikely that high moisture demanding crops will control high water table conditions in the high canal seepage area and or rapid water table rise during rainy season. Bio-drainage can however be effective as a complimentary system with other control methods. Their use for crop raising does not seems practical options as a substitute of subsurface drainage.

4. Preventive measures

Prevention is always better than cure is an old saying that has a direct relevance to waterlogging and soil salinity problems. In areas where waterlogging and salinity have not developed yet, it is necessary to take preventive measures to avoid soil degradation. Basically preventive measures can be grouped in three categories.

(i) Lining of water distribution system.

(ii) Improved on farm water management practices

(iii) Maintenance of surface drainage system.

Optimising water use efficiency on the farm is a priority objective for all irrigation projects throughout the world. Water use efficiency in developing countries like India may go as low as 30-40%. The most common on farm water losses includes surface runoff, deep percolation and evaporation. Improving on farm water management efficiency with good field conditions will save huge amount of water. For example proper design and layout of surface irrigation methods such as check basin and border may save irrigation water by 30 to 50%. Avoiding high water demanding crops and introduction of advanced irrigation methods like sprinkler and drip irrigation will be also useful in minimising losses. A complete package of improved on farm water management practices will be quite useful in combating waterlogging problems.

1.13 Subsurface Drainage Design Formula used in Different Countries

Transient and steady state subsurface horizontal drainage theories are being used in different parts of the world. Area receiving high rainfall (humid and sub-humid regions) over long period of time may permit use of steady state drainage theories, however in low rainfall area (arid and semi-arid regions) non-steady state or transient drainage theories for calculating drain spacing may be satisfactorily used. **Table 1.10** shows the drainage design equations used in different parts of the world.

Table 1.10 Drainage design equations used in different parts of the world

Country	Equation used for design
Australia	Hooghoudt Under some situations Glover and Integrated Hooghoudt and Hooghoudt-Ernst
Canada	Hooghoudt
Czechoslovakia	Ernst and Hooghoudt
Egypt	Hooghoudt
France	Non-steady state equation
Germany	Hooghoudt-Ernst
Hungary	Hooghoudt and Glover-Dumm
Iraq	Hooghoudt
Ireland	Kirkham, Hoohghoudt and Ernst
Japan	Kirkham
Jordan	Donnan or USBR method
The Netherlands	Hooghoudt and Ernst
Pakistan	Steady state equation, checked for transient flow conditions
Poland	Hooghoudt, Ernst and Kirkham
Romania	Hooghoudt and Ernst
U.S.A.	Donnan and USBR method

References

Abrol, I.P. and Bumbla, D.R. 1971. Saline and alkali soils in India; their occurrence and management FAO World soil Resource Rep, 41, pp 42-51.

Anonymous, 1996. Water and Related Statistics, Information System Directorate, Central Water Commission, Sewa Bhawan, New Delhi-110066.

Bhattacharya, A.K., 1999. Drainage of Agricultural Lands In: 50 Years of Natural Resource Management Research, edited by G.B. Singh and B.R. Sharma, an ICAR Publication, pp. 347-362.

Chauhan, H.S. and Sewa Ram, 1972. Surface and subsurface drainage hydrology. Proceeding of the Symposium on Waterlogging, Causes and Measures for It's Prevention. CBIP Publ. No. 118 pp. 19-26.

Chauhan, H.S. 2000. Development of Agricultural Drainage in the World. Proceeding of 8th International Drainage Workshop held at New Delhi from 31st January to 4th February, 2000, Vol. I, pp. 223-234.

INCID, 1991. Irrigated area in member countries, Central Office, New Delhi.

Inglis, C.S. and Gokhale, V.K.1928. Report on land drainage experiments at Baramati. Technical paper no. 24, Bombay PWD.

Johari, G.B., Maheshwari,K.M.,Schwab, G.O. and Bordurant B.L. 1990. Drainage of Irrigated Lands in India. Proceedings on Land drainage for salinity control in Arid and Semi Arid Regions February 25 March, 1990. Cario, Egypt.

Prasad, R.S.2000. Integration of drainage, flood control and water management in alluvial plains in India. Proceeding of 8th International Drainage Workshop held at New Delhi from 31st January to 4th February, 2000, Vol. I, pp. 33-66.

Rao, K.V.G.K. and Kamra, S.K.1983. Annual Report CSSRI, Karnal.

Richards, L.A. (Ed.) 1954. Diagnosis and Improvement of saline and alkali soils. Agri. Handbook Co, U.S.D.A., Washington D.C. 160 pp.

Sharma, B.R. and Paul, D.K.1999. Water Resources of India, In: 50 Years of Natural Resource Management Research, edited by G.B. Singh and B.R. Sharma, an ICAR Publication., pp. 31-48.

Singh, G.B. and, Sharma, B.R. (Ed.) 1999. 50 years of Natural Resource Management Research. Indian Council of Agricultural Research, pp 644.

Singh, N.T.1992. Land degradation and remedial measures with reference to salinity, alkalinity, waterlogging and acidity. In: Natural Resource Management for Sustainable Agriculture and Environment. De, D.L. (Ed.) Angkor Pub., New Delhi.

Tyagi, N.K. 1999. Management of salt affected soils In: 50 Years of Natural Resource Management Research, edited by G.B. Singh and B.R. Sharma, an ICAR Publication, pp.363-401.

Problems

1. What is waterlogging?
2. What are the different categories of waterlogging according to CGWB?
3. Describe the norms of waterlogging used by different states in the country.
4. Why there are different estimates of total waterlogged area in the country?
5. Describe common causes of waterlogging.
6. What are the different categories of salt affected soil?
7. Describe the need of drainage in the country.
8. What are the various disadvantages associated with poor drainage?
9. State various associated advantages of land drainage.
10. What are the different methods of drainage systems?
11. Write short notes on the following :
    a) Bio-drainage
    b) Vertical drainage
    c) Interceptor drainage
    d) Preventive measures for waterlogging
    e) Salt build-up in irrigated areas
    f) Extent of waterlogging in the country
    g) History of subsurface drainage in the country

Chapter 2

SATURATED HYDRAULIC CONDUCTIVITY

Determination of saturated hydraulic conductivity is of paramount importance for the design of mole drains, ditches, underground perforated pipe drains or well. Since ground water flow is under saturated conditions, therefore saturated hydraulic conductivity need to be determined prior to the design of these systems. Hydraulic conductivity varies very much from location to location and is moreover dependent on the direction of flow (anisotropy), temperature (viscosity) and some time period of year (swelling and shrinking of soil). Formula describing the flow of ground water, invariably assumes simplified conditions, whereby the hydraulic conductivity of certain idealized soil layers is considered constant. Hence a considerable difference exist between theory and practice. Therefore, selection of hydraulic conductivity for drainage design, out of number of observations taken from the field, is very important. Saturated hydraulic conductivity of the soil can be determined by using hydraulic method (based on Darcy's Law) or by correlation methods where the hydraulic conductivity of soil is found from such soil properties as pore size or grain size distribution, texture and structure etc. In the present chapter only most common hydraulic method will be discussed.

2.1 Laboratory Method

Method of determination of hydraulic conductivity in laboratory is based on soil samples, contained in cylinders of known diameters. Undisturbed soil samples must be taken in order to get a representative value of hydraulic conductivity. Stainless steel cylinders with a thin wall and one sharp end are used to extract soil samples above the ground water table. They are pressed gradually and evenly into the face of a profile pit. Care should be taken to avoid soil compaction. The soil around the cylinder is then removed and soil sample is withdrawn. The end of the sample should not be cut with a knife but should be removed wisely to expose the natural structure of the soil.

Undisturbed samples from the zone under the ground water table can be obtained by using a coring apparatus in a borehole. Closing of the tube above and below the samples with inflatable rubber rings, prevent the loss of material during extraction. The samples are only taken in vertical direction. **Figure 2.1** shows the apparatus required for the measurement of hydraulic conductivity. Hydraulic conductivity can be measured either by maintaining constant head or falling head (Wit, 1967).

2.1.1 Constant head method

The principle of the constant head method for a Kopecky cylinder is shown in **Figure 2.1**. Position of ring sample in the apparatus is shown in **Figure 2.2**. The induced steady state flow can be measured in a burette.

According to Darcy's law, K can be expressed as,

$$K = \frac{QL}{A.\Delta h} \tag{2.1}$$

where,

Q = steady state flow, $[L^3/T]$

Δh= constant head loss, [L]

A = cross sectional area, $[L^2]$

L = length of sample, [L]

Example 2.1

The sample in the Kopecky ring is 5.1 cm long and has a cross sectional area of 19.6 cm^2. The constant head loss maintained is 1.0 cm and the constant discharge rate is 190 cm^3/day. Find out hydraulic conductivity.

Solution 2.1

Substituting the above values in Equation (2.1) one will get.

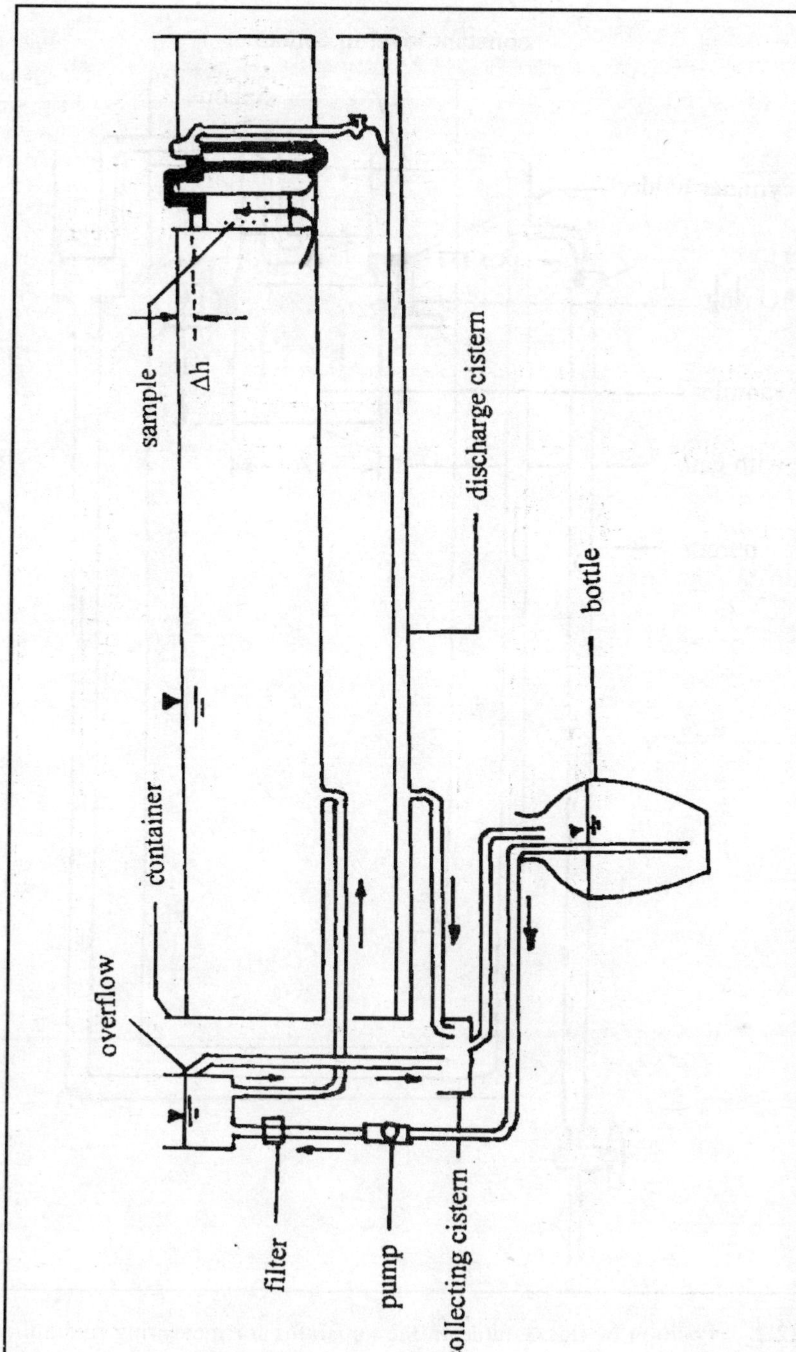

Figure 2.1. Apparatus for measuring hydraulic conductivity of undisturbed soil sample (Wit, 1967).

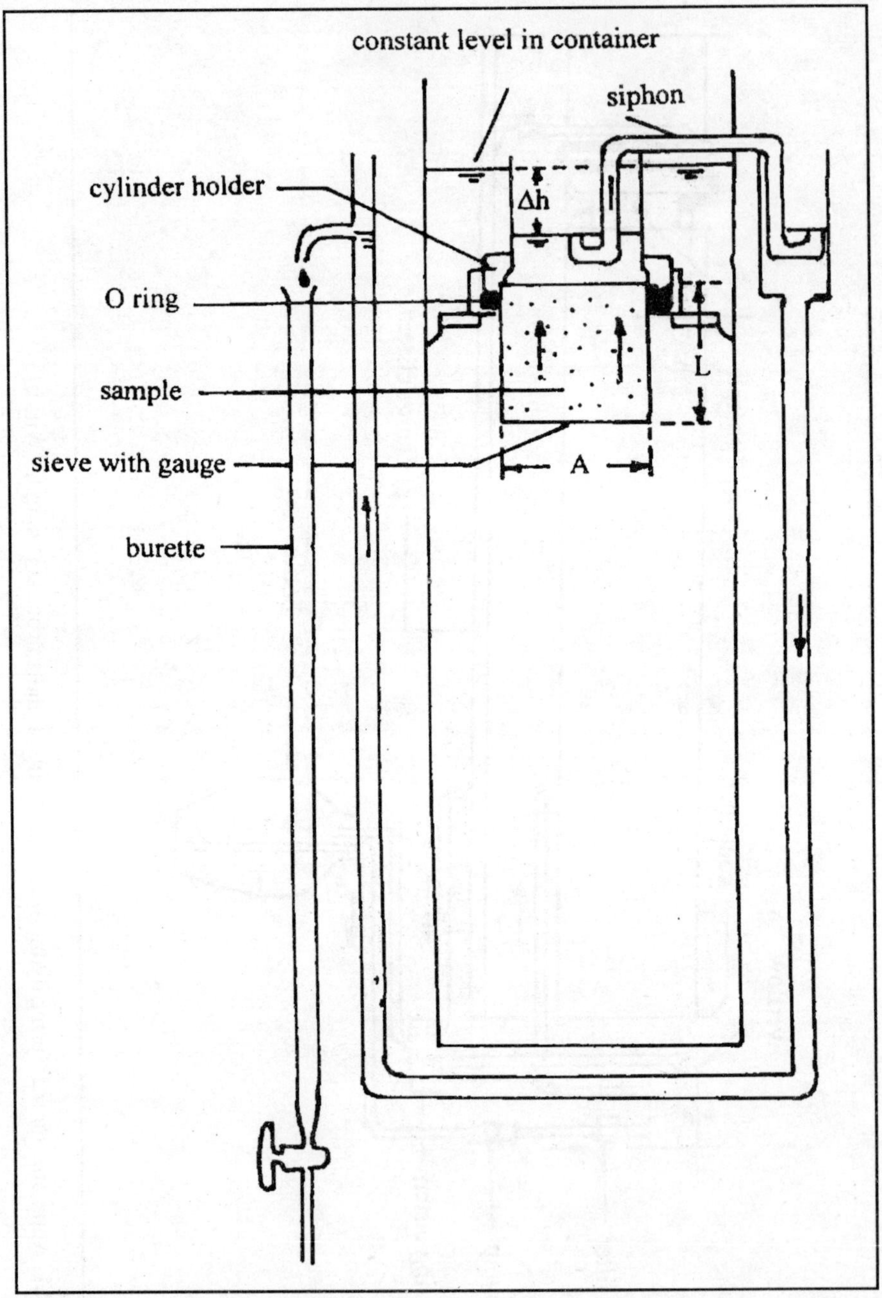

Figure 2.2. Position of ring sample in the apparatus for measuring hydraulic conductivity.

$$K = \frac{190X5.1}{1.0X19.6}$$

= 49 cm/day = 0.49 m/day

2.1.2 Falling head method

Figure 2.3 shows the principle of falling-head method for hydraulic conductivity measurement. A sample taken in a tube 6.2 cm in diameter and 30 cm height is placed in the container. After the water level above the sample reaches equilibrium, and the soil is saturated, the water on top is removed and consequent difference in head causes an upward flow through the sample. As a result the initial head loss and the flow are reduced. Measurements of the rate at which the head loss is reduced are used to calculate the hydraulic conductivity. If change in water head 'dh' taken in time 'dt' through the cross sectional area A, then the discharge or flow rate 'Q' can be expressed as below.

$$Q = -A_1 \frac{dh}{dt} \tag{2.2}$$

where,

A_1 = cross sectional area of observation tube in which water level changes, [L^2]

dh = the change in water level head in time, dt, [L]

dt = time taken to drop water level in a magnitude of, dh, [T]

Q = water flow rate, [L^3/T]

Negative sign shows the falling water level condition.

Q can also be estimated by using Darcy's law.

$$Q = -K.A.\frac{dh}{dl} \tag{2.3}$$

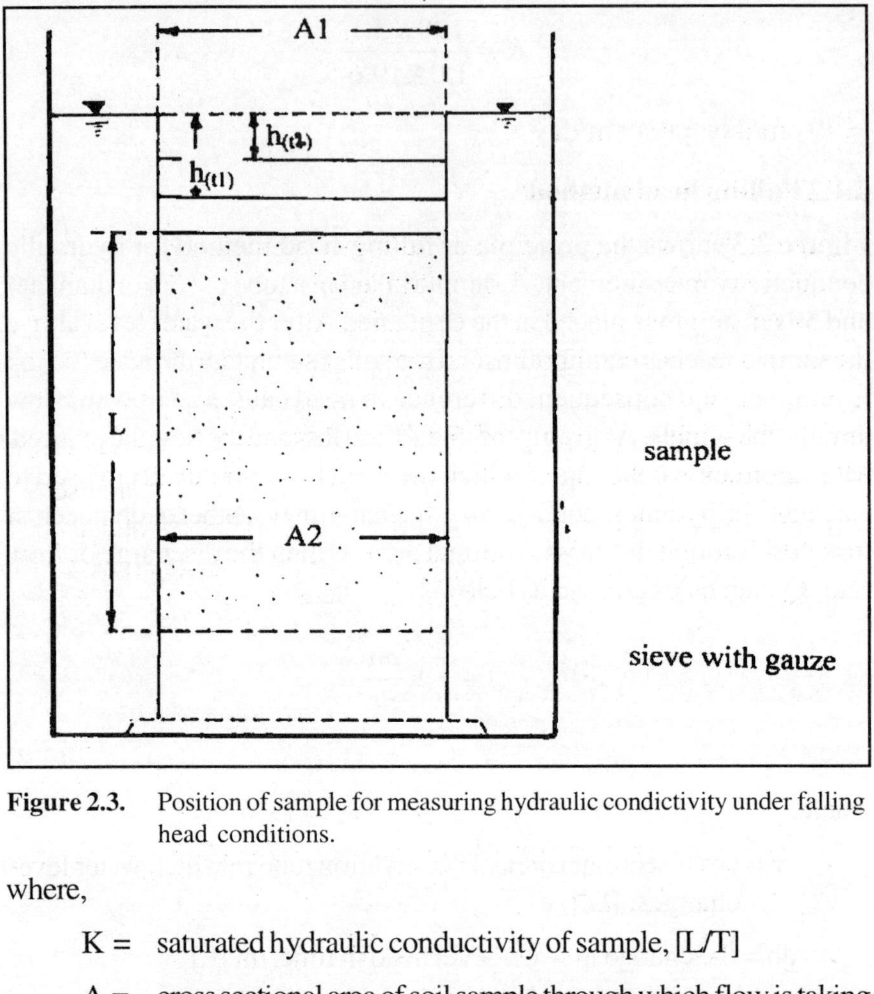

Figure 2.3. Position of sample for measuring hydraulic condictivity under falling head conditions.

where,

K = saturated hydraulic conductivity of sample, [L/T]

A = cross sectional area of soil sample through which flow is taking place, [L^2]

dh/dl= hydraulic gradient

If cross sectional area of the soil sample is A_2, length of the soil sample is L and water level head at any instant t is h then flow rate can be calculated using Equation (2.3).

$$Q = K.A_2 \frac{h}{L} \tag{2.4}$$

Combining Equation (2.2) and (2.4) will give,

$$-A_1 \frac{dh}{dt} = KA_2 \frac{h}{L} \tag{2.5}$$

Separating the variables and integrating under the limit, at $t = t_1$, $h = h_{t1}$ and at $t = t_2$, $h = h_{t2}$.

$$-\int_{h_{t1}}^{h_{t2}} \frac{dh}{h} = \frac{KA_2}{LA_1} \int_{t_1}^{t_2} dt \tag{2.6}$$

Solving above equation as under,

$$-[\ln h_{t_2} - \ln h_{t_1}] = \frac{KA_2}{LA_1}[t_2 - t_1]$$

$$\ln \frac{h_{t_1}}{h_{t_2}} = \frac{K}{L}\frac{A_2}{A_1}(t_2 - t_1)$$

$$K = \frac{A_1 L}{A_2 (t_2 - t_1)} \ln \frac{h_{t_1}}{h_{t_2}} \tag{2.7}$$

Example 2.2

A soil sample is taken below the ground water table in a zinc tube 6.2 cm in diameter and 30 cm long. After some soil is removed from the top and bottom, a soil sample of 25 cm remains. If observation tube has the same diameter as soil sample, find out hydraulic conductivity. It is also given that initial water level in measuring tube is 1.5 cm which after 7 hour reaches a level of 0.9 cm.

Solution 2.2

Given,

$$A_1 = A_2$$

At $t_1 = 0$, $h_{t1} = 1.5$ cm

At $t_2 = 7$ hour, $h_{t2} = 0.9$ cm

Substituting above values in Equation (2.7),

$$K = \frac{25}{\frac{(7-0)}{24}} X \ln\frac{1.5}{0.9} = 43, cm/day$$

2.2 Small Scale Field Method

Saturated hydraulic conductivity can be determined by small scale field methods both above and below a water table. When saturated hydraulic conductivity is to be determined above water table, inverse auger hole method is employed and when conductivity is to be determined below the water table, auger hole method has to be employed. For the above water table method one has to carry along a considerable amount of water because one must try to obtain a nearly saturated soil around the hole where one measures the hydraulic conductivity. With the below water table method one removes water from the cavity, so a pump or bailer is required but water has not to be provided for saturation. The water that leaves the soil and enters into the cavity, acts to keep the soil pores at the cavity wall open so that the boundary conditions tend to be better here than with infiltrometer test.

Auger hole method is simplest and most commonly used for determination of hydraulic conductivity below water table. In unstable soil sidewall of the hole must be protected with filter or unprotected auger hole can be used. In unstable soil piezometer method is selected for hydraulic conductivity measurement. This method has got advantage of determining the K value of a specific soil layer below the water table.

It is often observed that soils, which have been saturated for a long time, show reduced hydraulic conductivity values. Therefore, above water table methods may produce higher values of hydraulic conductivity than would be the case under saturated conditions. The essence of the above water table method is that one tries to obtain hydraulic conductivity values of a soil where as yet no drainage problem occur, but where they are expected to come into existence in the near future. Both questions, will drainage problem develop and will K remain unaltered, render above water table measurement rather speculative. Nevertheless, one-day drainage engineer may be asked whether an area with rising water table will need drainage or not. To answer this question infiltration measurements may helpful. If the part of auger hole cave in or if one is interested to determine K of a specific layer, part of hole may be protected by non-perforated pipe. For this case inverse piezometer method may be used, however, this method is not well developed.

2.2.1 Auger hole method

The principle of the method is as follows. A hole is bored into the soil with an auger to a certain depth below water table. When the water in the hole reaches equilibrium with the groundwater, part of it, is removed. The ground water then begins to seep into the hole and the rate at which it rises is measured. The hydraulic conductivity of the soil is computed with a formula or graph describing the mutual relation between the rate of rise, the ground water conditions and the geometry of the hole. This method measures the average hydraulic conductivity of a soil column about 30 cm in radius and extending from the ground water table to about 20 cm below the bottom of the hole, or to relatively impermeable layer if it occurs within 20 cm of the bottom. The method can also be used to measure the hydraulic conductivity of two separate layers.

2.2.1.1 Homogeneous Soil

1. Hooghoudt single auger hole method

Hooghoudt (1936) considered the first case that of homogeneous soil having no stratification and uniform hydraulic conductivity. The auger hole may or may not reach an impervious layer. One of the assumptions made in this derivation is that water table is not lowered around auger hole when water is bailed out of it. The condition is satisfied approximately for

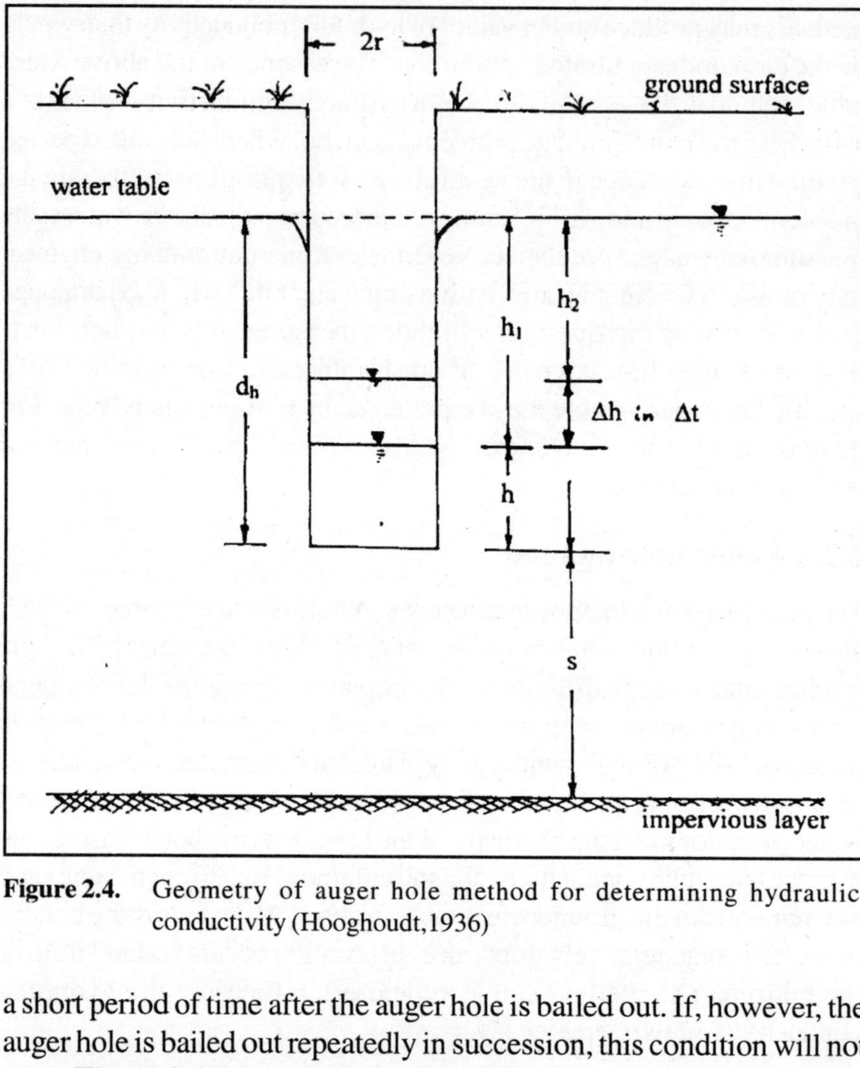

Figure 2.4. Geometry of auger hole method for determining hydraulic conductivity (Hooghoudt,1936)

a short period of time after the auger hole is bailed out. If, however, the auger hole is bailed out repeatedly in succession, this condition will not meet. **Figure 2.4** shows the geometry of flow.

Another assumption made by Hooghoudt is that water flows horizontally into the side of the auger hole and vertically up through the bottom of the hole. Actually the path of flow of the water must be curvilinear. Despite, the assumptions, Hooghoudt's formula are believed to be correct to within 10 to 20% if the rate of rise measurements are made when hole is not more than one fourth full. Assuming that the water flows horizontally through the side of auger hole, the rate at which the hole fills with water is proportional to the circumference of the hole and inversely proportional to the cross sectional area of the hole. The rate of rise of water in hole due to circumferential flow, at time t can be expressed as,

$$\frac{dh}{dt} = -K\frac{2\pi r.d_h}{\pi r^2}.\frac{h}{S} \tag{2.8}$$

where,

S = a constant which depends on r, d_h, and s and height of water in the hole at the time of measurement. Hooghoudt based on the results of his sand tank model tests proposed an empirical relationship (S = rd/0.19, [L]) in metric unit, using meter as unit length to predict the value of S.

h = depth of water level from initial water level in auger hole after bailing, [L]

d_h = depth of auger hole below water table, [L]

r = radius of auger hole, [L]

s = depth to impervious layer below the bottom of auger hole, [L]

If measurements are made in units other than meters, then Equation (2.8) must be changed. The constant S should depend on r, d_h, and s and also on the height of water in hole at the time of measurement. Hooghoudt determined the value of S with the aid of a controlled experiment in a sand tank and found that S in his assumed equation depend on r, the radius of hole and distance from impermeable layer to the water table. The experimental conditions under which Hooghoudt determined S only approximated field conditions because of the finite size of the sand tank used. Hooghoudt suggests that the coefficient 0.19 is accurate to within 27% of true value, an accuracy which he considers adequate for determination of hydraulic conductivity which may vary in field from 0.001 to more than 10 m/day. The numerical coefficient 0.19 has the dimension of $L^{-1}$and is valid only for meter^{-1}. Approximate conversion factors are necessary to convert to other units of length.

Water also flows upward through the bottom of the hole and an approximate expression for rate of rise of water in the hole due to this flow was assumed to be.

$$\frac{dh}{dt} = -K\frac{\pi r^2}{\pi r^2}.\frac{h}{S} = -\frac{Kh}{S} \tag{2.9}$$

Adding Equation (2.8) and (2.9),

$$\frac{dh}{dt} = -\frac{2Kd_h h}{rS} - \frac{Kh}{S} = -K(\frac{2d_h + r}{rS})h \tag{2.10}$$

which is the rate of rise in the hole due to water entering both the circumference as well as the bottom of the hole. Integrating the Equation (2.10) between the limits h = h_1 at t = 0 and h = h_2 at t = t.

$$\int_{h_1}^{h_2} -\frac{dh}{h} = \int_0^t K\frac{(2d_h + r)}{rS}dt$$

$$-\ln h\Big]_{h_1}^{h_2} = K\frac{2d_h + r}{rS}.t\Big]_0^t$$

$$\ln\frac{h_1}{h_2} = K(\frac{2d_h + r}{rS})t$$

$$K = \frac{r.S}{(2d_h + r)t}\ln\frac{h_1}{h_2} \tag{2.11}$$

Introducing logarithm to the base 10,

$$K = \frac{2.3r.S}{(2d_h + r)t}\log_{10}\frac{h_1}{h_2} \tag{2.12}$$

When the auger hole terminates on impervious barrier the vertical flow of water through the bottom equals zero and Equation (2.12) becomes,

$$K = \frac{2.3r.S}{2d_h.t}\log_{10}\frac{h_1}{h_2} \tag{2.13}$$

Equation (2.12) can also be written as follows,

$$t = \frac{2.3r.S}{(2d_h + r)K}\log_{10}\frac{h_1}{h_2} \tag{2.14}$$

If time t is plotted on linear scale and (h_1/h_2) on log scale, one will get a straight line. From the slope (M) of plotted line hydraulic conductivity can be calculated as below.

$$K = \frac{2.3r.SM}{(2d_h + r)} \tag{2.15}$$

Example 2.3a

(I) Derive an expression for saturated hydraulic conductivity from Hooghoudt governing equation for the boundary conditions, t = 0, h = ho and t = t, h = h_t.

(II) For the measurement of saturated hydraulic conductivity an auger hole of 20 cm diameter was made up to a depth of 1.50 m below the ground surface. Initial water table depth below ground surface was measured to be 0.50 m. When water level in the hole reached the initial ground water level, the water from the hole was bailed out to a level of 90 cm below initial water level. Rise in water level was recorded with time as given below.

t, min.	0	10	20	30	40	50	60	70	80	90	100	110	120
h_t, cm	90	65	60	49	40	33	27	22	21	20	19	18.5	18

Calculate the saturated hydraulic conductivity in meter per day from above data set.

Solution 2.3a

(I) Governing equation of hooghoudt for flow towards auger hole can be written below.

$$\frac{dh}{dt} + K(\frac{2d_h + r}{rS})h = 0$$

Boundary conditions: $t = 0 \qquad h = h_o$

$t = t \qquad h = h_t$

Now separating the variables and integration upon given boundary conditions above equation may be rewritten as,

$$\int_{h_o}^{h_t} \frac{dh}{h} = -K(\frac{2d_h + r}{rS})\int_0^t dt$$

$$\ln h\Big]_{h_o}^{h_t} = -K(\frac{2d_h + r}{rS})t\Big]_0^t = \ln h_t - \ln h_o = -K(\frac{2d_h + r}{rS})(t - 0)$$

$$\ln\frac{h_t}{h_o} = -K(\frac{2d_h + r}{rS})t = \ln\frac{h_o}{h_t} = K(\frac{2d_h + r}{rS})t$$

$$K = \frac{rS}{(2d_h + r)t}\ln\frac{h_o}{h_t} = K = 2.303\frac{rS}{(2d_h + r)t}\log_{10}\frac{h_o}{h_t}$$

Above equation may be also written as,

$$Y = Mx + C$$

Where,

$Y = \ln (h_o / h_t)$

$M = \text{slope} = K\ [(2d_h+r)/(rS)]$

$C = 0 = \text{intercept}$

(II) Given:

$r = 0.20/2 = 0.10$ m

$d_h = (1.50 - 0.50)$ m $= 1.0$ m

$h_o = 0.90$ m

(1) Perform the following calculation,

Time, minute	h_t, cm	h_o/h_t	ln (h_o/h_t)
0	90.0	1.000	0.000
10	65.0	1.385	0.325
20	60.0	1.500	0.406
30	49.0	1.837	0.608
40	40.0	2.250	0.811
50	33.0	2.727	1.003
60	27.0	3.333	1.204
70	22.0	4.091	1.409
80	21.0	4.286	1.455
90	20.0	4.500	1.504
100	19.0	4.737	1.555
110	18.5	4.865	1.582
120	18.0	5.000	1.609

(2) Prepare a graph between ln (h_o/h_t) on y-axis and time t on x-axis. Measure the slope from the straight portion of the line as below (**Figure 2.5**).

$$M = (Y_2-Y_1)/(t_2 - t_1) = (1.20 - 0.60)/(60-30) = 0.60/30 = 0.02$$

Note: Slope of the line may be also worked out by plotting above data on semi-log paper [(h_o/h_t) on log scale and t on normal scale]. For the calculation of saturated hydraulic conductivity Equation (2.15) can be used.

(3) Calculate $S = rd_h/0.19 = 0.10 \times 1.0/0.19 = 0.526$ m

(4) Substitute the values of S, r, d_h and M in the expression given below to calculate the hydraulic conductivity.

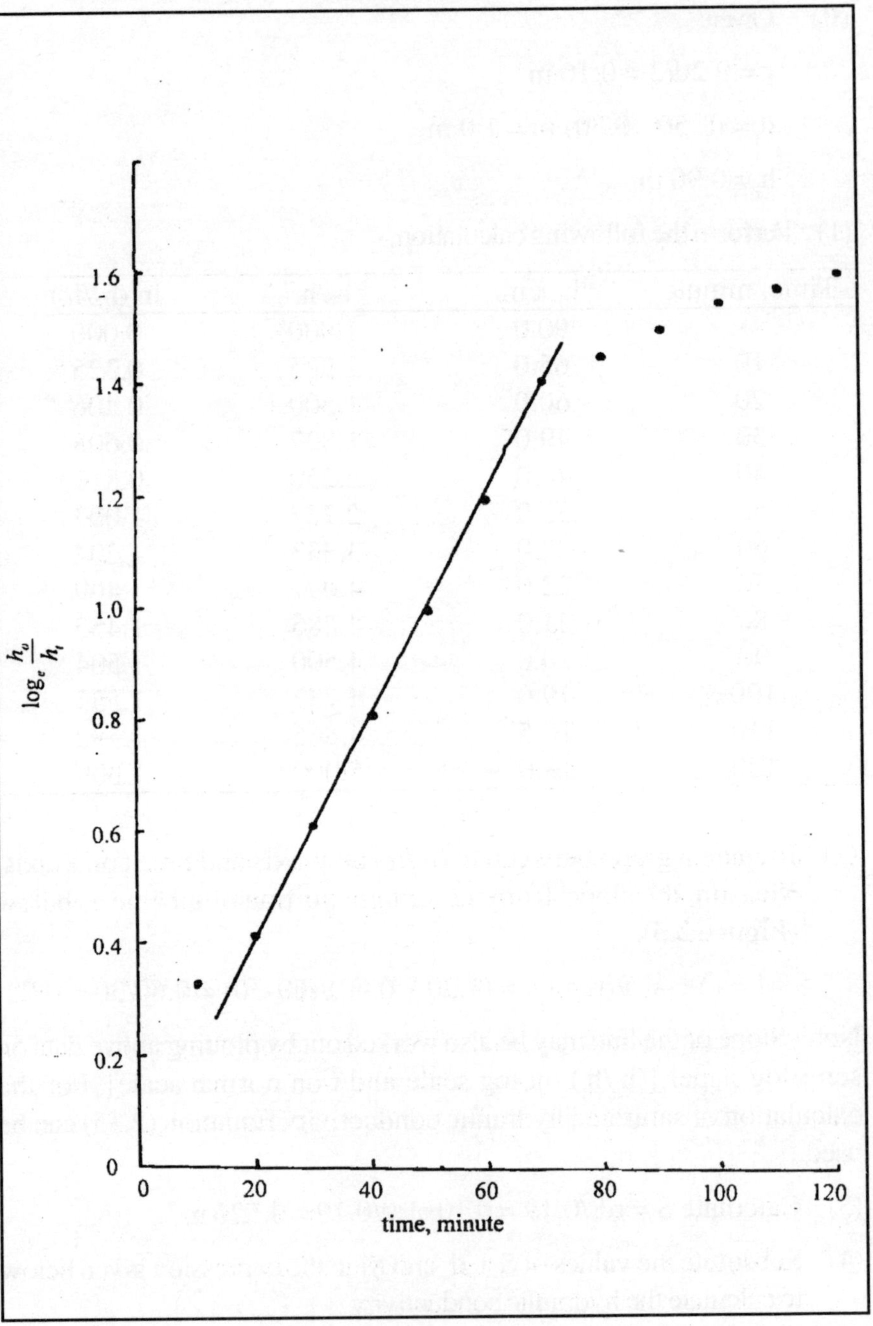

Figure 2.5. Relationship between $\log_3 (h_o/h_t)$ and time, t.

$$M = K\frac{(2d_h + r)}{rS}$$

$$K = \frac{rMS}{(2d_h + r)} = \frac{0.10x0.02x0.526}{2x1.0 + 0.10} = 5.0095X10^{-4} m/\min$$

$$= 0.72 m/day$$

Example 2.3 b

Calculate the saturated hydraulic conductivity of the soil using water levels in hole at time t = 50 minute and t = 120 minute for the Problem 2.3a and compare the calculated values of hydraulic conductivity with the earlier calculated value.

Solution 2.3 b

Given:

$r = 0.10$ m

$d_h = 1.0$ m

$h_o = 0.90$ m at t = 00 minute

$h_{50} = 0.33$ m at t = 50 minute

$h_{120} = 0.18$ m at t = 120 minute

$S = 0.526$ m

1. Substituting the data in expression for calculating the saturated hydraulic conductivity for t = 50 minute

$$K = \frac{rS}{(2d_h + r)t}\ln\frac{h_o}{h_t} = \frac{0.10x0.526}{(2x1.0 + 0.10)50}\ln\frac{h_o}{h_{50}}$$

$$= \frac{0.10x0.526}{(2x1.0 + 0.10)x50}\ln\frac{0.90}{0.33} = 5.0245x10^{-4} m/\min$$

$$= 0.72 m/day$$

2. Substituting the data in the expression for calculating the saturated hydraulic conductivity for t=120 minute

$$K = \frac{rS}{(2d_h + r)t} \ln \frac{h_o}{h_t} = \frac{0.10x0.526}{(2x1.0 + 0.10)x120} \ln \frac{h_o}{h_{120}}$$

$$= \frac{0.10x0.526}{(2x1.0 + 0.10)x120} \ln \frac{0.90}{0.18} = 3.359x10^{-4} m/\min$$

$$= 0.484 m/day$$

Comparison of these two estimates of hydraulic conductivity shows that the value calculated for t = 120 minute is lower than the value estimated for t = 50 minute.

Example 2.3c

An auger hole of 6 cm diameter was dug up to the impervious barrier to work out the saturated hydraulic conductivity in a waterlogged area. The hole was left undisturbed for 24 hours to coincide the water level in the hole with the initial water level of the area. Water level in the hole was brought to a level of 90 cm below the initial level by bailing out the water from the hole. If water level in the hole rises 10 cm in 19 second what will be the saturated hydraulic conductivity of the soil.

Solution 2.3c

Given: r = 0.06/2 = 0.03 m

h_1 = 0.90 m

h_2 = (0.97 – 10) m = 0.87 m

Δt = 19 second

1. Substitute the value of S = rd/0/19 in Equation (2.13) as below.

$$K = \frac{2.3r.rd_h}{2d_h \Delta t.0.19} \log_{10} \frac{h_1}{h_2}$$

$$= \frac{2.3r^2}{0.38} \frac{\log_{10} \frac{h_1}{h_2}}{\Delta t}$$

2. Substitute the values of r, h_1, h_2 and Dt in above equation.

$$K = \frac{2.3x0.03^2}{0.38} \cdot \frac{\log_{10} \frac{0.90}{0.87}}{19}$$

$$= 4.222x10^{-6} m / \sec$$

$$= 0.365 m / day$$

Discussion

An unlined hole of 7.5 to 15 centimeter diameter extending below the water table is generally suitable for taking observations provided the soil strata is suitable and the wall of the hole is stable. In case of unstable soil layers and under quick sand conditions the side walls of the hole should be lined with a perforated pipe. Care should be taken to avoid compaction of the walls of the auger holes. After digging out the hole to desired depth, it is left for several hours till the water level reaches its static water table height. Observations of rising water levels in the hole should be taken till the water surface rises to about one third of the drawdown. Some time it is necessary to repeat the entire operation two or more times to get an accurate representative value of hydraulic conductivity. Use of single value for determining drain spacing is not recommended, hence several measurements should be taken preferably making a grid. An appropriate technique should be employed to select the representative value of saturated hydraulic conductivity for drainage design.

2. Ernst single auger hole method

Ernst (1950) assumed the rate of rise of water level in the hole is proportional to the K value of the soil and the difference in height of water level outside and inside the hole. Further it depends on the radius of the hole. He developed an emperical graphical numerical solution of this flow problem. **Figure 2.5** shows the geometry of the hole and equation is written below.

$$K = C\frac{\Delta h}{\Delta t} \qquad (2.16)$$

where

$$C = \frac{4000}{(20+\frac{d_h}{r})(2-\frac{\bar{h}}{d_h})}\frac{r}{\bar{h}}$$

when $s > 1/2\ d_h$, (2.17)

$$C = \frac{3600}{(10+\frac{d_h}{r})(2-\frac{\bar{h}}{d_h})}\frac{r}{\bar{h}}$$

when s=0

Where,

C = geometry factor f (h, d_h, r, s)

$\Delta h/\Delta t$ = rate of rise of water table in the auger hole, [L/T]

$\Delta h = h_o - h_t$, h_o and h_t are the depth (in cm) of the water level in the hole at time t_0 and t (sec) after bailing out water, respectively.

Δt = t – to, (in sec)

$h = (h_o + h_t)/2 = h_o - (1/2)\ \Delta h$

d_h = depth of auger hole below water table, [L]

s = depth of impermeable layer below the bottom of the hole, [L]

The geometry factor C can be determined from **Figure 2.7** when $s > d_h/2$ and from **Figure 2.8** when s = 0.

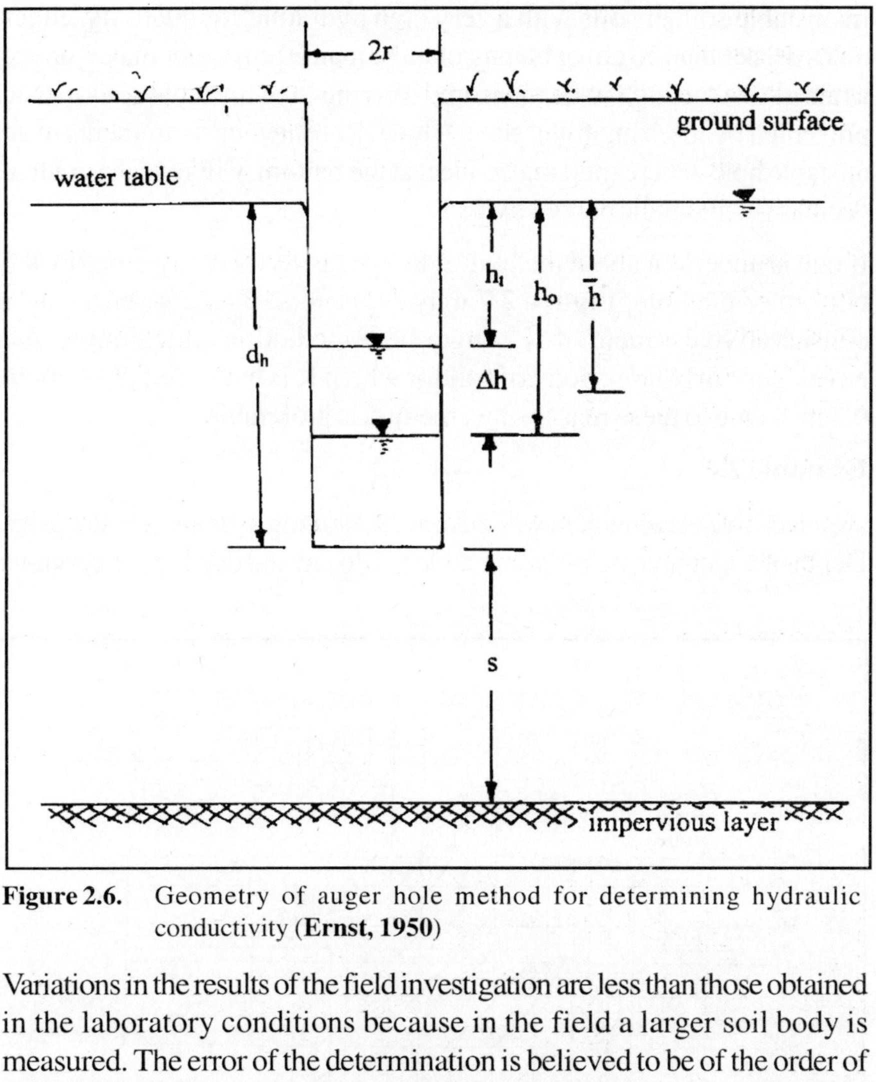

Figure 2.6. Geometry of auger hole method for determining hydraulic conductivity (**Ernst, 1950**)

Variations in the results of the field investigation are less than those obtained in the laboratory conditions because in the field a larger soil body is measured. The error of the determination is believed to be of the order of 10 to 20%. In stable soils of low hydraulic conductivity, it may be necessary to extend the auger hole 1 m or more below the water table, and to bail out more than 80 cm of water to create a measurable rate of rise. In heavy soils the action of auger hole may destroy the soil structure in the vicinity of auger wall of hole. It will then be necessary to flush the soil and reopen the closed soil pores by bailing the water out several times. Repeated measurements will reveal that the hydraulic conductivity gradually improve with time.

In unstable soil, or soils with a very high hydraulic conductivity, auger holes deeper than 50 cm or bailing of more than 20 cm water may produce a rate of rise too high to be measured. Caving in of unstable soils can be prevented by inserting filter into the hole. Repeated measurements in an unstable hole, where mud may collect at the bottom, will reveal a gradual decrease in hydraulic conductivity.

If one is uncertain about the hydraulic conductivity of layer below the bottom of the hole, **Figure 2.7** may be used ($s>d_h/2$), for ensuring a conservative estimate of K. Auger hole method is quite simple and envelopes fairly large body of soil for which K is measured (0.05 m^3 to 0.3 m^3). Due to these reasons this method is preferable.

Example 2.4

An auger hole of radius 4cm was made for measuring hydraulic conductivity. Depth of auger hole below water table is 126 cm and depth of impervious

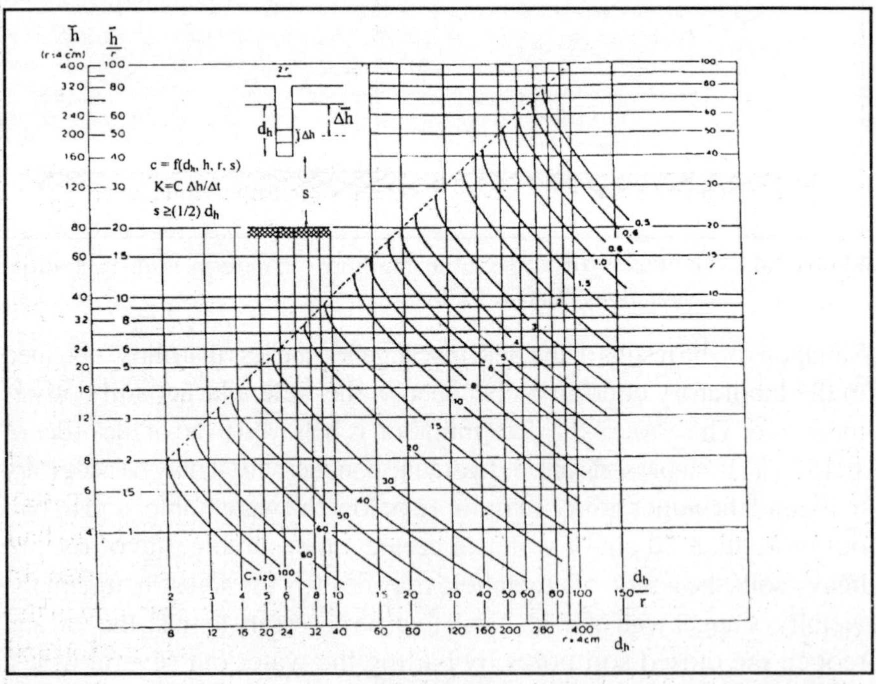

Figure 2.7. Nomograph for determining C in augur hole metod of **Ernst (1950)** when $s>d_h$

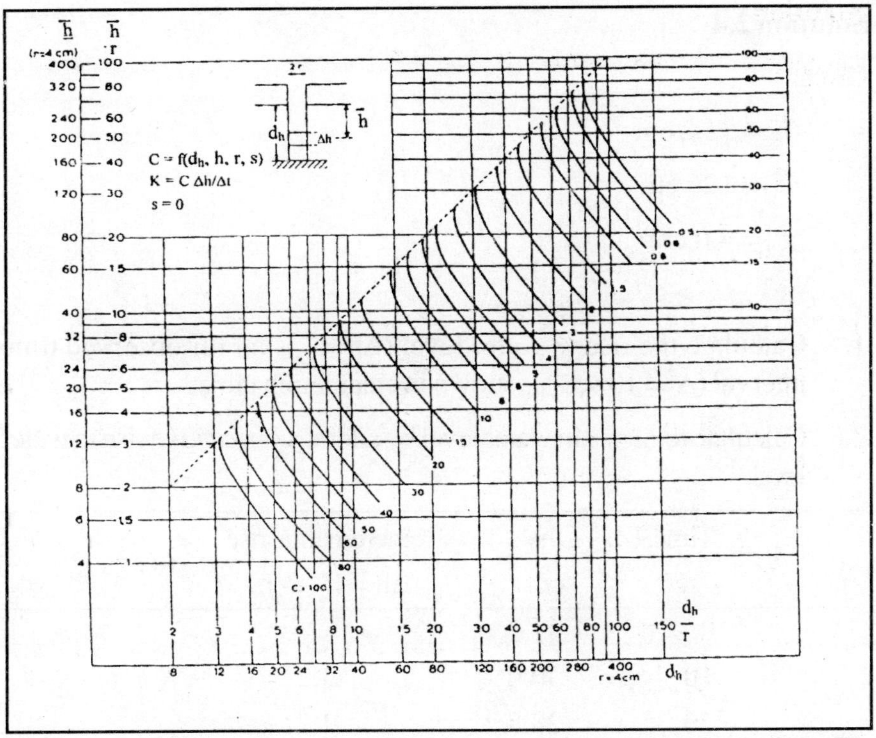

Figure 2.8. Nomograph for determining C in auger hole method of **Ernst (1950)** when s = 0.

layer below the bottom of hole is 150 cm. Level of water in the hole immediately after bailing out water is 31.2 cm. Work out saturated hydraulic conductivity of the soil if the rise of water level with respect to time is as below.

time, sec	h_t, cm
0	31.2
10	30.0
20	28.8
30	27.7
40	26.6
50	25.6

Solution 2.4

Given:

$h_o = 31.2$ cm

$d_h = 126$ cm,

$r = 4.0$ cm

$s_{(=150\ cm)} > d_h/2_{\ (=126/2 = 63\ cm)}$

1. Calculate the rise in water level ($\Delta h = h_{t-1} - h_t$) for observed time interval ($\Delta t = 10$ sec) as shown in calculation table.
2. Calculate total rise in water level by adding consecutive rise in water level,

Time, t sec	h_t cm	incremental rise $\Delta h = h_{t-1} - h_t$
0	31.2	-
10	30.0	1.2
20	28.8	1.2
30	27.7	1.1
40	26.6	1.1
50	25.6	1.0

Total rise = 1.2+1.2+1.1+1.1+1.0 = 5.6 cm.

3. Calculate $\bar{h} = (h_o+h_t)/2 = [31.2 + (31.2 - \text{total rise})]/2 = [31.2 + 25.6]/2 = 28.4$ cm.
4. Check $\Delta h < ho/4 \Rightarrow 5.6 < 31.2/4$ (=7.8). This condition is satisfied.
5. Calculate $d_h/r = 126 \div 4 = 31.5$ and $\bar{h}/r = 28.4 \div 4 = 7.1$.
6. Check $s > d_h/2 \Rightarrow 150 > 126/2$, select **Figure 2.7** for obtaining the value of C.
7. Move 7.1 unit vertically upward on y-axis and 31.5 unit in horizontal direction parallel to x-axis and find out value of correction factor C = 6.0.

8. Using Equation (2.17) calculate value of K.

   K = C (Δh/Δt) = 6.0 (5.6/50) = 0.672 m/day.

3. Piezometer method

The principle of this method is similar to single auger hole method, except that a tube is inserted into the hole to leave a small cavity at the bottom of the hole. The method is also called as pipe cavity method.

After augering a hole into the soil to desired depth below the water table, a pipe is inserted into the hole leaving a small unprotected cavity at the bottom. Care should be taken to avoid leakage along the wall of the hole. After water level in the pipe reaches equilibrium with ground water, the water from pipe is removed. Due to sudden pressure difference water starts flowing from the surrounding soil into the uncased cavity. The rate of rise is measured and hydraulic conductivity is calculated using appropriate formula describing the corresponding flow conditions.

Kirkham (1945) gave a general formula for flow into the cavity below piezometers. Flow geometry for this case is shown in **Figure 2.9**.

(2.18)

$$\frac{dQ}{dt} = -K.c.h$$

If water level dh rise in time dt then,

$$Q = \pi r_p^{\ 2}.dh$$

(2.19)

where,

r_p = radius of the pipe, [L]

Q = volumetric flow of water, $[L^3]$

c = f(d_h, r_c, l, s), geometry factor

r_c = radius of uncased cavity, [L]

l = length of uncased cavity, [L]

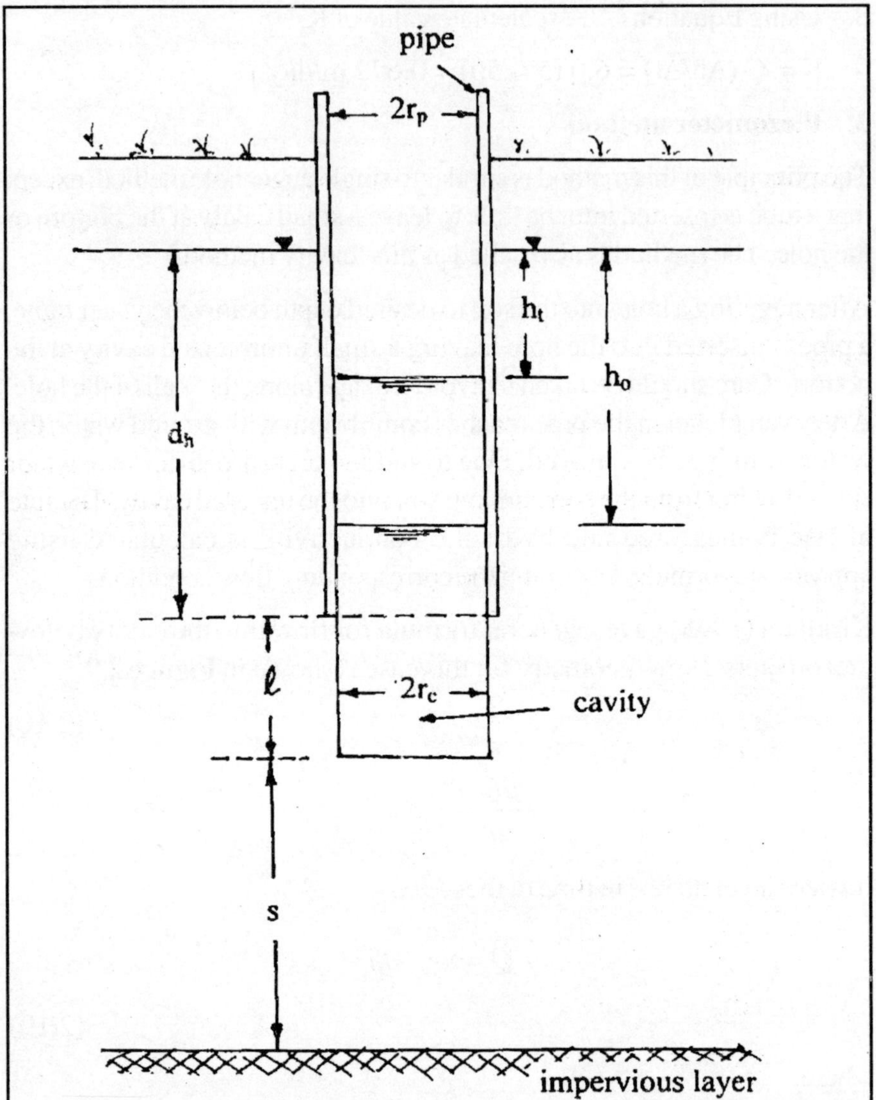

Figure 2.9. Piezometer method for determining hydraulic conductivity (**Luthin and Kirkham, 1949)**

s = depth of impervious layer below the bottom of the cavity, L]

h = depth to water level below static water table, [L]

Combining Equation (2.18) with Equation (2.19) will yield,

$$-\pi r_p^2 \frac{dh}{dt} = cK.h$$

$$-\frac{dh}{h} = \frac{c.K.}{\pi r_p^2} dt$$

(2.20)

Integrating above equation under the limit, at $t = t_o$, $h = h_o$ and $t = t$, $h = h_t$ one will obtain.

$$\ln \frac{h_o}{h_t} = \frac{c.K(t - t_o)}{\pi r_p^2}$$

$$K = \frac{\pi r_p^2 \ln \frac{h_o}{h_t}}{c(t - t_o)}$$

(2.21)

The geometry factor C was determined with an electric analogue, whereas in the auger hole method it was determined by relaxation. **Figure 2.10** may be used to estimate the value of geometry factor. The piezometre method is suitable for determining the hydraulic conductivity of individual soil layers or of deep layer. Eighty per cent of hydraulic head is dissipated in the soil within a distancc of one cavity length from the cavity and about 90 per cent within a distance of two cavity lengths. This method measures the hydraulic conductivity of the soil within a radius of one or two cavity lengths from the cavity (0.01 to 0.05 m^3), in a horizontal direction. In unstable soil cavity must be supported. The smaller inflow surface of cavity compared with that provided by the auger hole method, increases the entrance velocity of the water. This increases the risk of collapse or deformation of cavity and might lead to considerable error. If the wall of the cavity is sealed by augering action in stable soil, it must be flushed several times for good result.

Example 2.5

A piezometer with an inside radius of 5 cm is driven into the soil to 1 m below the water table. A cavity of 25 cm deep and 5 cm in radius was made below the piezometer. After equilibrium is reached, water was pumped out of the piezometer. Following measurements of the rate of rise are recorded.

Time, sec	0	20	40	60	80
Depth of water level after pumping, cm	75	65	60	56	52

Determine the hydraulic conductivity of the soil.

Solution 2.5

Given:

$r_p = 5$ cm

$d_h = 100$ cm

$l = 25$ cm

1. Calculate the ratio between $\{\ln(h_o/h_t)\}/(t-t_o)$ for all reading to check out the consistency in the data series.

t, sec	h_t, cm	$\ln(ht/h_o)/(t-t_o)$
0	75	-
20	65	$3.68\text{x}10^{-3}$
40	60	$3.85\text{x}10^{-3}$
60	56	$3.72\text{x}10^{-3}$
80	52	$3.71\text{x}10^{-3}$

2. Determine the ratio of $l/r_c = 25/5 = 5$ and $d_h/r_c = 100/5 = 20$

3. Move 5 unit on x-axis and 20 unit vertically upward parallel to y-axis and read the geometry factor c from **Figure 2.10** as = 100.

4. Calculate hydraulic conductivity by using Equation (2.21) in cm/sec, and to convert it into m/day multiply it by a factor 864 as given below.

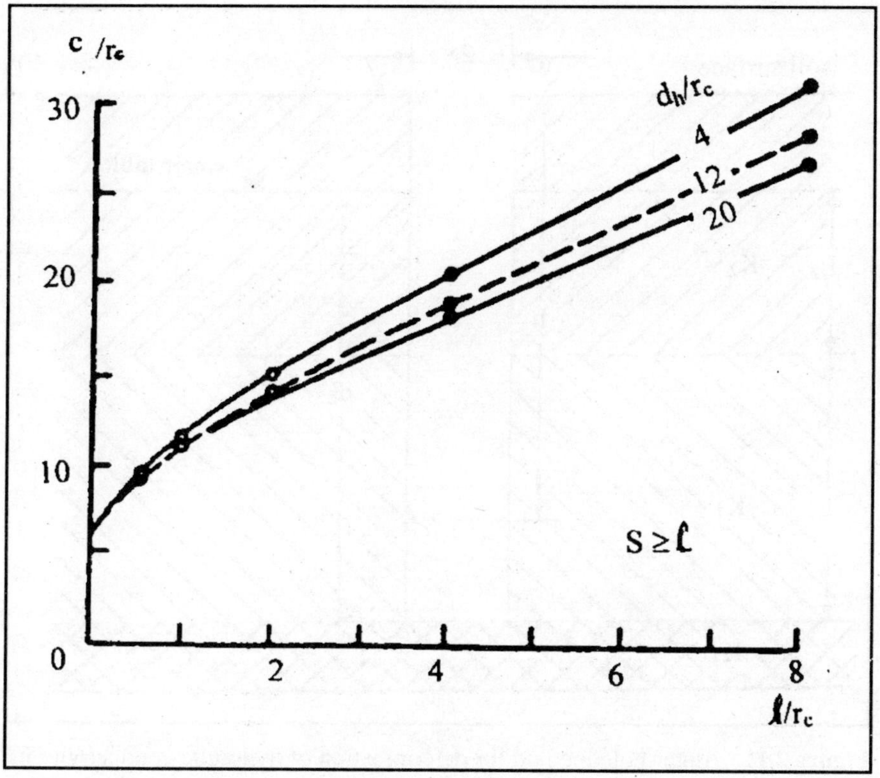

Figure 2.10. Nomograph for determination of C in piezometer method.

$$K = \frac{864 X \pi X r_p^2}{c(t - t_o)} \ln \frac{h_o}{h_t} = \frac{864 X \pi X 5^2}{100} X 3.71 X 10^{-3}$$

$= 2.5$ m/day

2.2.1.2 Layered Soil

1. Hooghoudt single auger hole method

Consider a soil profile consisting of two sharply divided layers as shown in **Figure 2.11**. A auger hole is extended into the second layer and the rate of rise of water in the auger hole due to water seeping out of first layer at time t is assumed to be given by,

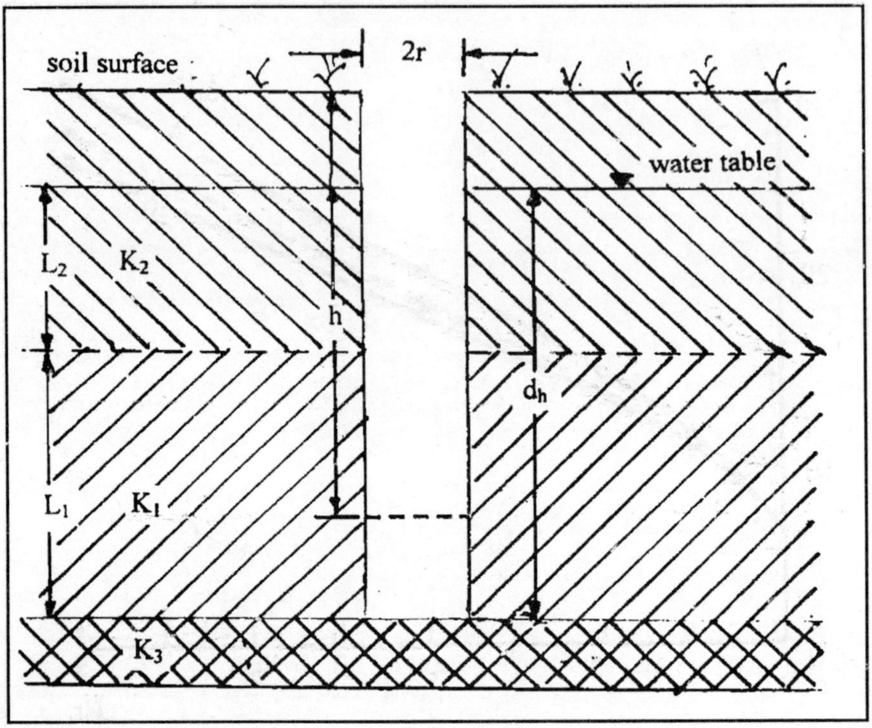

Figure 2.11. Auger hole method for determination of hydraulic conductivity in layered soil (**Hooghoudt,** 1936).

$$\frac{dh}{dt} = -K_1 \frac{2\pi r L_1}{\pi r^2} \frac{h}{S} = -K_1 \frac{2L_1 h}{r.S} \tag{2.22}$$

where,

K_1= saturated hydraulic conductivity of first layer, [L/T]

L_1= thickness of the first layer, [L]

Other notations are similar to the Hooghoudt single auger hole for homogeneous soil. The assumptions made are that all the streamlines are horizontal and the hydraulic head loss across soil layer is numerically equal to the distance between the water table and the water level in the hole. The contribution due to second layer is assumed on the same basis to be given by,

$$\frac{dh}{dt} = -K_2 \frac{2\pi r L_2}{\pi r^2} \frac{h}{S} = -K_2 \frac{2L_2 h}{r.S} \tag{2.23}$$

where,

K_2 = saturated hydraulic conductivity of second layer, [L/T]

L_2= thickness of second soil layer below water table to the interface, [L]

Rate of rise of water level in auger hole due to vertical flow from the bottom of the hole can be expressed as below.

$$\frac{dh}{dt} = -K_3 \frac{\pi r^2}{\pi r^2} \frac{h}{S} = -K_3 \frac{h}{S} \tag{2.24}$$

Where,

K_3 = saturated hydraulic conductivity of third or bottom layer, [L]

Adding Equations (2.22), (2.23) and (2.24) and regrouping the terms and integrating from $h = h_1$ to $h = h_2$ and $t = 0$ to $t = t$.

$$\frac{dh}{dt} = -(\frac{2K_1L_1 + 2K_2L_2 + K_3 r}{rS})h \tag{2.25}$$

$$-\int_{h_1}^{h_2} \frac{dh}{h} = \frac{2K_1L_1 + 2K_2L_2 + K_3 r}{rS} \int_0^t dt$$

$$\ln \frac{h_1}{h_2} = \frac{2K_1L_1 + 2K_2L_2 + K_3 r}{rS} t \tag{2.26}$$

Changing the natural logarithm to a logarithm to the base 10, which yields.

$$\frac{\log_{10}\dfrac{h_1}{h_2}}{t} = \frac{2K_1L_1 + 2K_2L_2 + K_3r}{2.3rS} \tag{2.27}$$

If the hole terminates on the impervious layer then Equation (2.27) reduces to,

$$\frac{\log_{10}\dfrac{h_1}{h_2}}{t} = \frac{2K_1L_1 + 2K_2L_2}{2.3rS} \tag{2.28}$$

For upper layer of soil hydraulic conductivity is first determined as discussed for the case of homogeneous soil and then for second and third layer by lowering the same hole in second and third layer, respectively.

For the case of three layered soils with the auger hole not terminating on an impervious layer, the formula is,

$$\frac{\log_{10}\dfrac{h_1}{h_2}}{t} = \frac{2K_1L_1 + 2K_2L_2 + K_3L_3 + K_4r}{2.3rS} \tag{2.29}$$

Where,

K_4 = saturated hydraulic conductivity of fourth layer, [L/T]

L_4 = thickness of fourth soil layer, [L]

2. Ernst two layered soil

If a soil profile consists of two layers of different hydraulic conductivity, K_1 in the upper layer and K_2 in the lower layer, these values for separate layers can be determined if the water table is well within the upper layer.

Two successive measurements are made in one hole. First measurement is taken when hole is above the interface and the second measurement is taken after original hole has been deepened into the lower layer below the interface. First a hole is bored at least 40 cm below the water table, but should not extend farther than 20 cm above the lower layer. The deepened bored hole must reach at least 50 cm into the lower layer. Flow geometry is shown in **Figure 2.12**. The measurement in the shallow hole gives K_1, the hydraulic conductivity of upper layer in the same way as for homogeneous soil profile.

$$K_1 = C_1 \frac{\Delta h}{\Delta t} \tag{2.30}$$

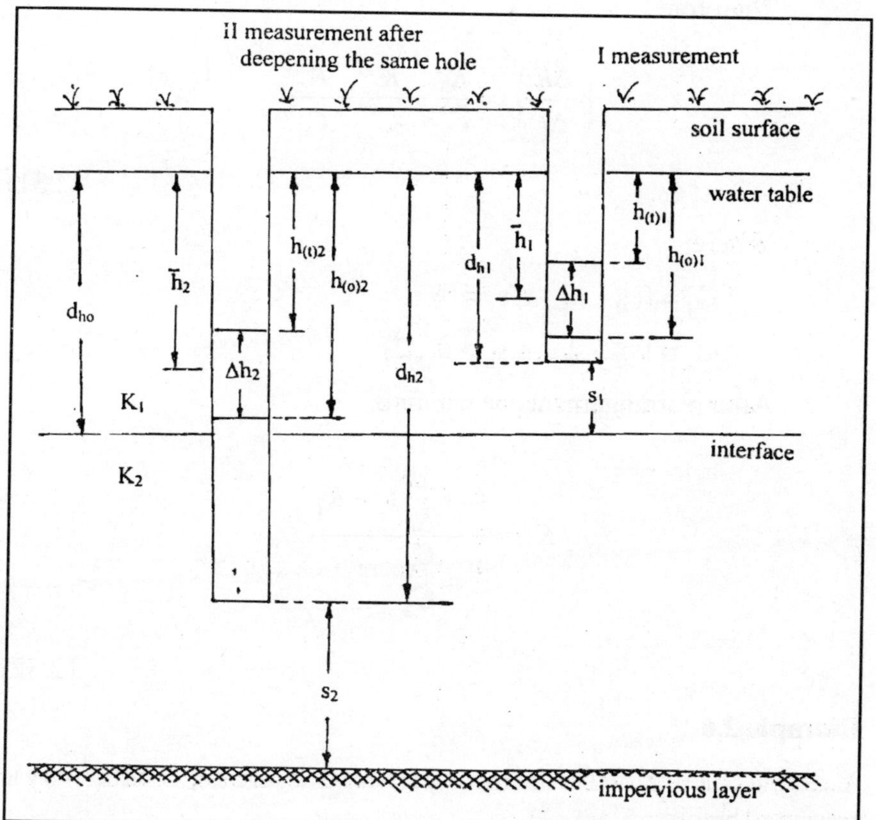

Figure 2.12. Auger hole method for determination of hydraulic conductivity in two layered soil (Ernst, 1950).

Where,

$C_1 = f(\bar{h}_1, d_{h1}, r, s_1 > d_{h1}/2)$

The rate at which the water in the deep hole rises is thought to be the result of two components.

- Inflow of water from the upper layer having hydraulic conductivity, K_1 only and lower layer is considered impervious.
- Inflow of water from the lower layer which is considered to be having inflow from whole profile with hydraulic conductivity K_2 minus inflow from the upper layer also with hydraulic conductivity K_2, the lower layer being considered impervious.

Therefore,

$$\left(\frac{\Delta h}{\Delta t}\right)_2 = \frac{K_1}{C_o} + \frac{K_2}{C_2} - \frac{K_2}{C_o} \tag{2.31}$$

Where,

$C_o = f(\bar{h}_2, d_{ho}, r, s_1 = 0)$

$C_2 = f(\bar{h}_2, d_{h2}, r, s_2 > d_{h2}/2)$

After rearrangement one will find.

$$K_2 = \frac{C_o\left(\frac{\Delta h}{\Delta t}\right)_2 - K_1}{\frac{C_o}{C_2} - 1} \tag{2.32}$$

Example 2.6

A sample calculation for the measurement of hydraulic conductivity is presented below.

$(\Delta h/\Delta t)_1$	=	0.16
d_{h1}	=	70 cm
$\bar{h}_1$	=	30 cm
$s_1 > d_h$		
r	=	4 cm

Read the value of C_1 from **Figure 2.6**, which comes out to be 9.4.

$$K_1 = C_1 (\Delta h/\Delta t)_1 = 9.4 \times 0.16 = 1.5 \text{ m/day}$$

$(\Delta h/\Delta t)_2$	=	0.26	d_{ho}	=	100 cm
d_{h2}	=	150 cm	h_2	=	40 cm
$\bar{h}_2$	=	30 cm	s_1	=	• 0
$s_2 > d_h$					
r	=	4 cm			

Read the value of C_2 from **Figure 2.7**, which comes out to be 3.9.

Read the value of Co from **Figure 2.8**, which comes out to be 6.3.

$K_2=\{C_o(\Delta h/\Delta t)_2-K_1\}/(C_o/C_2-1)=\{(6.3 \times 0.26-1.5)/(6.3/3.9-1)\}/(6.3/3.9-1) =$

2.2.2 Inverse auger hole method

This method is employed to determine saturated hydraulic conductivity of the soil in absence of water table. This method is based on the principle of infiltration: the velocity of infiltration (v) equals the saturated hydraulic conductivity, (K). If one bore a hole into the soil and fills it with water until the soil around the hole and below the hole is practically saturated, the velocity of infiltration becomes more or less constant. The total infiltration (Q) will then be equal to,

$$Q = v.\ A \tag{2.33}$$

Where,

A = surface area of infiltration, $[L^2]$

Since the velocity of infiltration, v = K.i, where, i is the hydraulic gradient and may be treated as unity, therefore, v becomes equal to K and Equation 2.33 takes the following form.

$$Q = K.A \tag{2.34}$$

For the inverse auger hole method, infiltration occurs both through the bottom and side walls of the holes (**Figure 2.13**). If the radius of hole is r and depth of water in the hole is h, the total area of water flow becomes, $A = (\pi r^2 + 2\pi.r.h)$. Substituting the value of A in Equation (2.34) will give us,

$$Q = 2\pi Kr(h + \frac{1}{2}r) \tag{2.35}$$

Further we can find Q from the rate of fall of water level in the hole as, $Q = \pi r^2 (dh/dt)$. Substituting this value of Q in above equation will yield.

$$2\pi Kr(h + \frac{1}{2}r) = -\pi r^2 \frac{dh}{dt}$$

$$2K(h + \frac{1}{2}r) = -r\frac{dh}{dt} \tag{2.36}$$

Upon integration over the limit, at $t = t_o$, $h = h_o$ and at $t = t$, $h = h_t$

$$-\int_{h_o}^{h_t} \frac{dh}{(h + \frac{1}{2}r)} = \frac{2K}{r}\int_{t_o}^{t} dt$$

$$\ln\frac{(h_o + \frac{1}{2}r)}{(h_t + \frac{1}{2}r)} = \frac{2K}{r}(t - t_o)$$

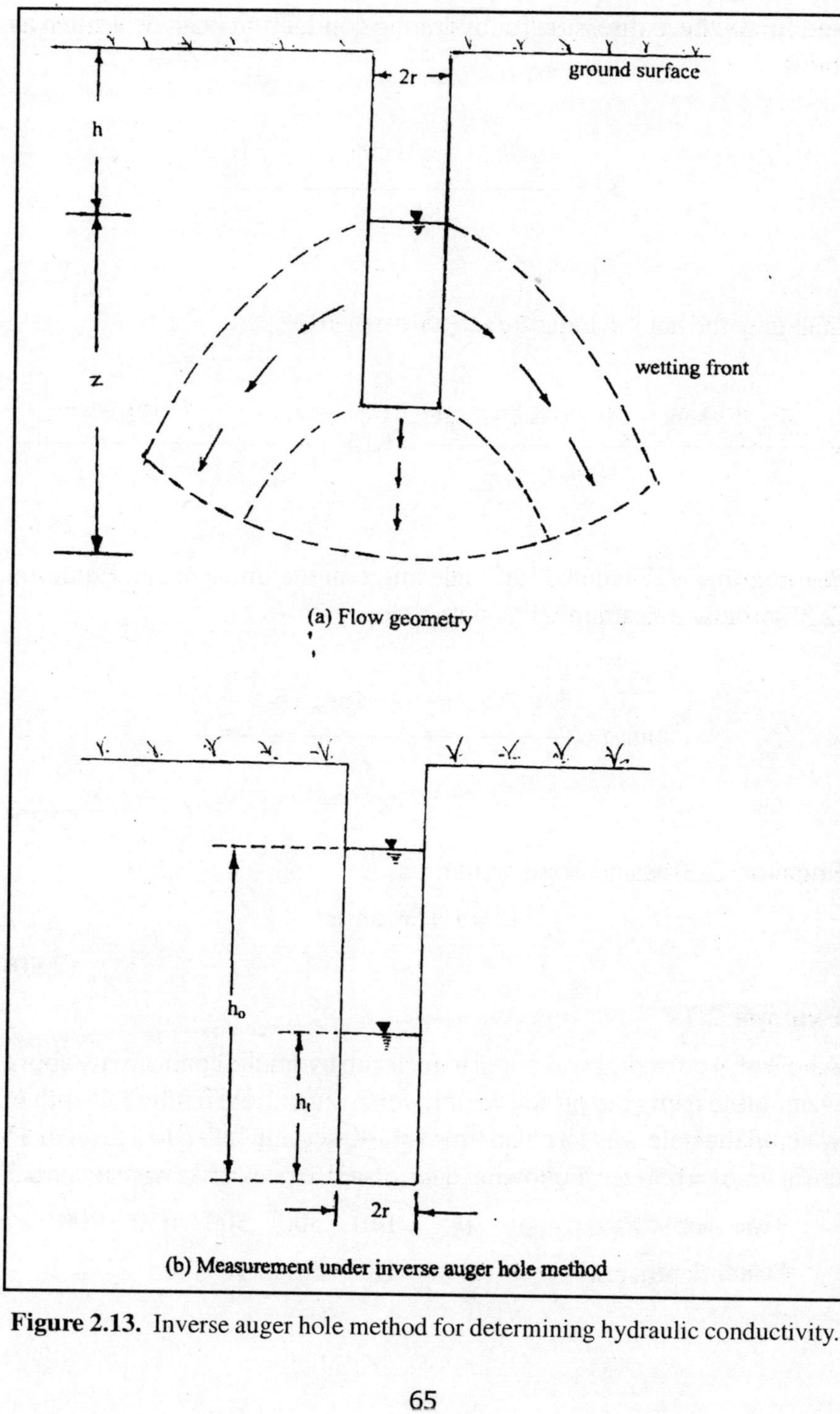

Figure 2.13. Inverse auger hole method for determining hydraulic conductivity.

And finally the expression for hydraulic conductivity can be written as below.

$$K = \frac{r \ln(h_o + \frac{1}{2}r) - \ln(h_t + \frac{1}{2}r)}{2(t - t_o)}$$

(2.37)

Changing the natural logarithm to a base of 10.

$$K = \frac{2.3r}{2} \frac{\log_{10}(h_o + \frac{1}{2}r) - \log_{10}(h_t + \frac{1}{2}r)}{(t - t_o)} = 1.15r \frac{\log_{10}(h_o + \frac{1}{2}r) - \log_{10}(h_t + \frac{1}{2}r)}{(t - t_o)}$$

(2.38)

Plotting (h_t+1/2 r) on the log scale and t on the linear scale, Equation (2.38) produces a straight line with a slope,

$$\tan\alpha = \frac{\log_{10}(h_0 + \frac{1}{2}r) - \log_{10}(h_t + \frac{1}{2}r)}{(t - t_o)}$$

(2.39)

Equation (2.38) can also be written as,

$$K = 1.15r \tan\alpha$$

(2.40)

Example 2.7

A hole of 4 cm radius was bored to measure hydraulic conductivity above water table using the principle of inverse auger hole method. Depth of water in the hole was 18 cm at time t = 140 sec and fallen to a level of 15 cm at time t = 650 sec. Following data related to rate of rise were recorded.

Time, sec	0	140	300	500	650	900
Water depth, cm	19	18	17	16	15	14

Determine hydraulic conductivity of the soil layer located above water table.

Solution 2.7

Given: $r = 4$ cm

$h_o = 18$ cm $t_o = 140$ sec

$t_0 = 15$ cm $t = 630$ sec

1. Calculate $h_t + r/2$ for all the time.
2. Take $\log_{10}$ of $(h_t + r/2)$ of respective values,

time, sec	h_t	$h_t + (1/2)r$	$\log_{10}[h_t + (1/2)r]$
0	19	21	1.322
140	18	20	1.301
300	17	19	1.279
500	16	18	1.255
650	15	17	1.231
900	14	16	1.204

3. Plot $(h_t + r/2)$ on log scale or $\log(h_t + r/2)$ on normal scale (y-scale) and time t on linear scale (x-scale) [**Figure 2.14**],
4. Draw a best fit straight line with visual observation.
5. Calculate slope of line from the plotted line or by regression analysis but consider only those data, which are following linear relationship. Data point lying far away from the line should be left out. This may happen towards the end of line.
6. Calculate K-value as follow,

$t_o = 140$ sec $\quad h_o+(1/2)r = 20 \quad \log_{10} h_o+(1/2)r = 1.30$

$t = 650$ sec $\quad h_t+(1/2)r = 17 \quad \log_{10} h_t+(1/2)r = 1.23$

$$K = 1.15 X 4 X \frac{1.30 - 1.23}{650 - 140}$$

$= 0.00063$ cm/sec $=0.55$ m/day

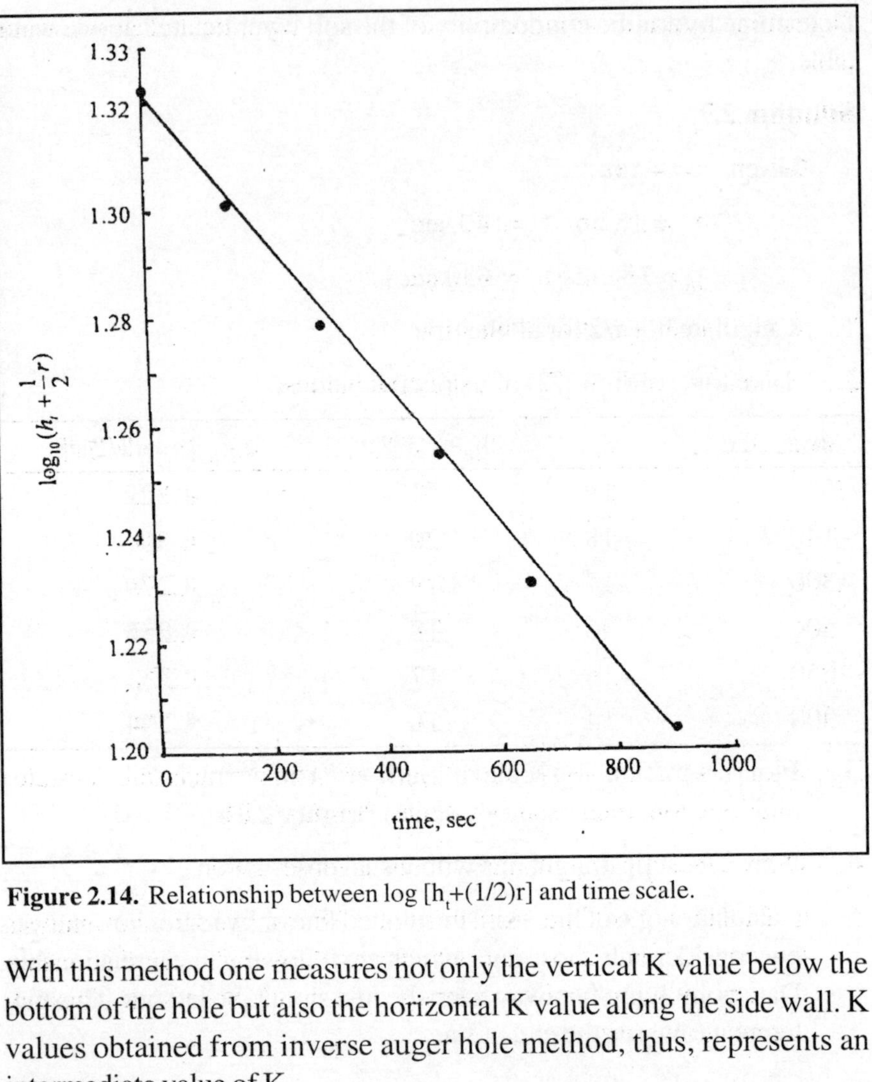

Figure 2.14. Relationship between log [h_t+(1/2)r] and time scale.

With this method one measures not only the vertical K value below the bottom of the hole but also the horizontal K value along the side wall. K values obtained from inverse auger hole method, thus, represents an intermediate value of K.

References

Beers, W.F.J. van. 1965. The augerhole method. Bull. No. 1. ILRI, Wageningen, The Netherlands.

Bouwer, H. and R.D. Jackson. 1974. Determining soil properties. In: Drainage for Agriculture, (J.M. van Schilfgaarde, edited), Am. Soc. of Agronomy, USA.

Ernst, L.F.1950. A new formula for the calculation of the permeability factor with the auger hole method. T.N.O. Groningen. Translated from Dutch by H. Bouwer, Cornell Univ., Ithaca, N.Y.1955.

Hooghoudt, S.B.1936. Bijdragen tot de kennis van enige natuurkundige groothenden van den grond. Verslagen Landbouwkundig Onderzoek, 42(13)B:449-541. The Hague

Hooghoudt, S.B.1937. Contributions to the knowledge of certain physical properties of the soil. Brondenkundig Institute, Croningen, The Netherlands, Bulletin No. 43(1)B:461-676.

Hoorn, J.W. van.1979. Determining hydraulic conductivity with the inverse auger hole and infiltration methods. In: Proceedings of the International Drainage Workshop. (J. Wesseling, ed.)Publ.25, ILRI, Wageningen, The Netherlands.

Kirkham, D. and CHM van Bavel.1948. Theory of seepage into augerholes. Proceedings of the Soil Science Society of America, 13:75-81.

Kirkham, D.1945. Proposed method for field measurement of permeability of soil below the water table. Proceedings of the Soil Soil Science Society of America, 11:93-99.

Luthin, J.N.1957. Drainage of Agricultural Land. American Society for Agronomy, Madison, Wisconsin, USA.

Smiles, D.E. and EG. Young.1965. Hydraulic conductivity determination by several field methods in a sand tank. Soil Science, 99:83-87.

Wit, K.E. 1-967. Apparatus for measuring hydraulic conductivity of undisturbed soil samples. Technical Bulletin 52. Institute for Land and Water Management Research, Wageningen, 12, p.

Wit, K.E.1962. An apparatus for coring undisturbed samples in deep bore holes. Journal of Soil Science.94:65-70.

Young, E.G.1968. Shape factor for Kirkham's piezometer method for determining the hydraulic conductivity of soil in situ for soils overlying an impermeable floor or inifinity permeable stratum.Soil Science, 106:235-237.

Problems

Theoretical

1. Describe two laboratory methods of hydraulic conductivity measurement and their limitations.
2. Dervie an expression for calculation of hydraulic conductivity for falling head method under laboratory conditions.
3. Dervie an expression for calculating saturated hydraulic conductivity for Hooghoudt's auger hole method. Also state the assumptions involved in the derivation of the expression and it's limitations.
4. Write the expression for the calculations of hydraulic conductivity for Ernst's single auger hold method when the hole rests well above the impervious layer and when it terminates at impervious layer.
5. Derive an expressin for calculating saturated hydraulic in absence of water table using an inverse auger hole techique.

Numerical

1. A soil sample was taken below the ground water table in a zinc tube of 6.0 cm diameter and 35 cm length. Some soil was removed from both side of the sample in such a way that a soil sample of 24.5 cm length remains. If the dimater of observation tube and soil sample is the same, calculate the hydraulic conductivity of the sample. Initial water level which was measured 2.0 cm initially, came down to a level of 0.8 cm after 8 hour.

   [Answer: 67.3 cm/day]

2. A 5.0 cm long soil sample having a cross sectional area of 20 cm2 was removed from the experimental site,carefully. Water was allowed to flow under a constanthead of 1.5 cm and measured discharge rate was 190.8 cm3/day. Calculate hydraulic conductivity of the soil sample.

   [Answer: 31.8 cm/day]

3. An auger hole of 10 cm diameter was dug out upto the impervious barrier. Auger hole was left undisturbed for 48 hours to coincide the

water level in hole with initial water level of the area. The depth of hold was 1.20 m. Water level in the hole was brought to a level of 100 cm below the initial level by bailing out water from the hole. I the water level rise 5 cm in 25 second, what will be the saturated hydraulic conduct ivity of the soil.

[Answer: 3.69 cm/day]

4. An auger hole of 0.20 cm diameter was constructed in such a way that the depth of hole below static water level becomes 1.0 m. Water from the hole was bailed out and observations of rising water level were taken against time. The log values of (ho/ht) was plotted against time (minute) and the slope of linear portion of the line was worked out to be 0.02 minute-1. Calculate the hydraulic conductivity of the soil by using Hooghoudt single auger hole method. If the dimaeter of the hole is doubled using Hooghoudt single auger hole method. If the diameter of the hole is doubled keeping other measurements same, what will now be the new hydraulic conductivity?

[Answer: 0.72 cm/day, 1.38 m/day]

5. An auger hole of 0.30 cm diameter was constructed in a water logged area in such a way that the depth of auger hole extends to a level of 1.2 m below static water table. Immediately after bailing out water from the hole water level reaches a level 0.80 m and after 120 minute it rises to a level of 0.16 m. Calculate the hydraulic condutivity of the soil.

[Answer: 1.08 cm/day]

6. For the measurement of saturated hydraulic conductivity an auger hole of 8 cm diameter was constructed. The depth of hole below the static water level was 120 cm and depth to impervious layer from the bottom of the hole was 120 cm.; after bailing out of water from the hold, following observations were recorded.

Time, second	ht, cm
00	50.0
20	47.1
40	44.5
60	42.1
80	39.7
100	37.7

Calculate hydraulic conductivity of the soil suign Ernst single auger hole approach.

[Answer: 0.57 cm/day]

6. For the measurement of hydraulic conductivity in absence of water level an auger hole of 6 cm diameter was constructed. The water was repeatedly filled in the hole to saturate the soil as around the hole. Following observations were recorded after filling the hole with water :-

time, second	0	150	300	450	600	750
height of water in hole above the bottom, cm	33	28	23	18	16	13

Calculate the hydraulic conductivity of the soil.

[Answer: 3.21 cm/day]

Chapter 3

STEADY AND TRANSIENT DRAIN SPACING FORMULAS

3.1 Types of Drains

Systems of open ditches were in use since long to control water table. In modern agriculture, pipe drains are replacing the systems of open ditches. Concrete and clay tiles have been used in several field drainage systems successfully. Availability of perforated pipes has replaced the conventional materials used for pipe drainage. Cost of the drainage system depends on the spacing between tile/pipe lines. Drains are of different types and shown in **Figure 3.1.**

a) **Field Drains:** They are also known as field laterals or simply laterals. These laterals are usually parallel drains whose function is to control ground water table positions. They collect water from the surrounding soil mass and convey it to the collector drains. Water seeps into field drains through the perforation made on the pipes or through the joints of the tiles or from both.

b) **Collector Drains:** The main function of collector drains is to collect water from individual field drains and dispose it off into the main drains. It also controls ground water positions to some extent.

c) **Main Drains :** The function of main drains is to collect water from collector drains and transport it out of the drainage area. It controls ground water position, as well.

3.2 Factors Influencing Midway Water Table and Assumptions Involved

Spacing between the field drains are very important as it decides the cost

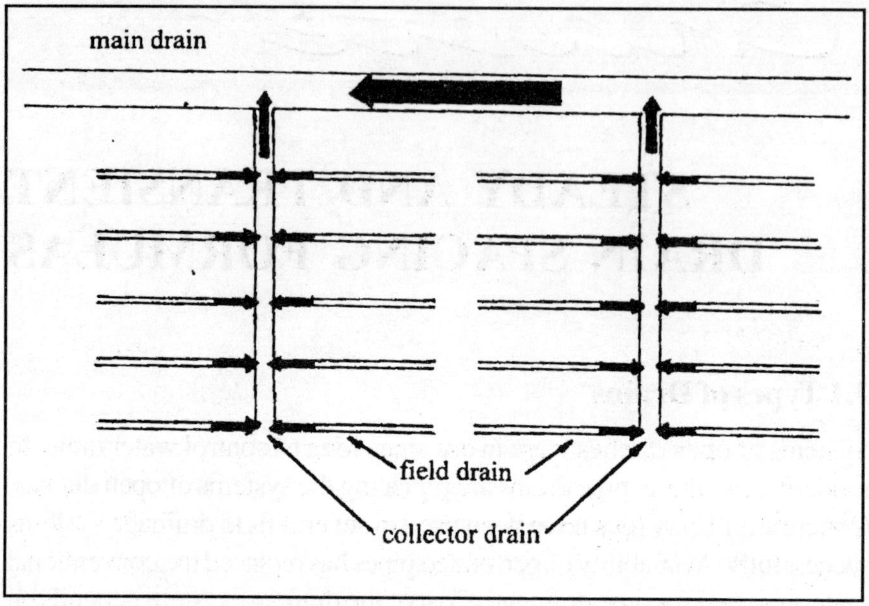

Figure 3.1. Function of different types of drains.

of the system and the efficiency and efficacy of the installed system. All the drain spacing formulas are derived for parallel field ditches/drains. **Figure 3.2** shows a cross sectional view of the laterals. Water table between two parallel field drains has curved shape with highest position midway between the drains. Following factors influence the midway water table height which is critical for crops.

1. Rainfall and different sources of recharge e.g. surface runoff, seepage from canals and upland areas etc.
3. Evaporation and different sources of discharge e.g. ground water pumping and flow.
3. Cross sectional area of the ditch/drain.
4. Water level in the ditches or at the drain.

These factors are inter-related by drainage equations, based on the following assumptions :

- The flow is identical in any cross section perpendicular to the drains.

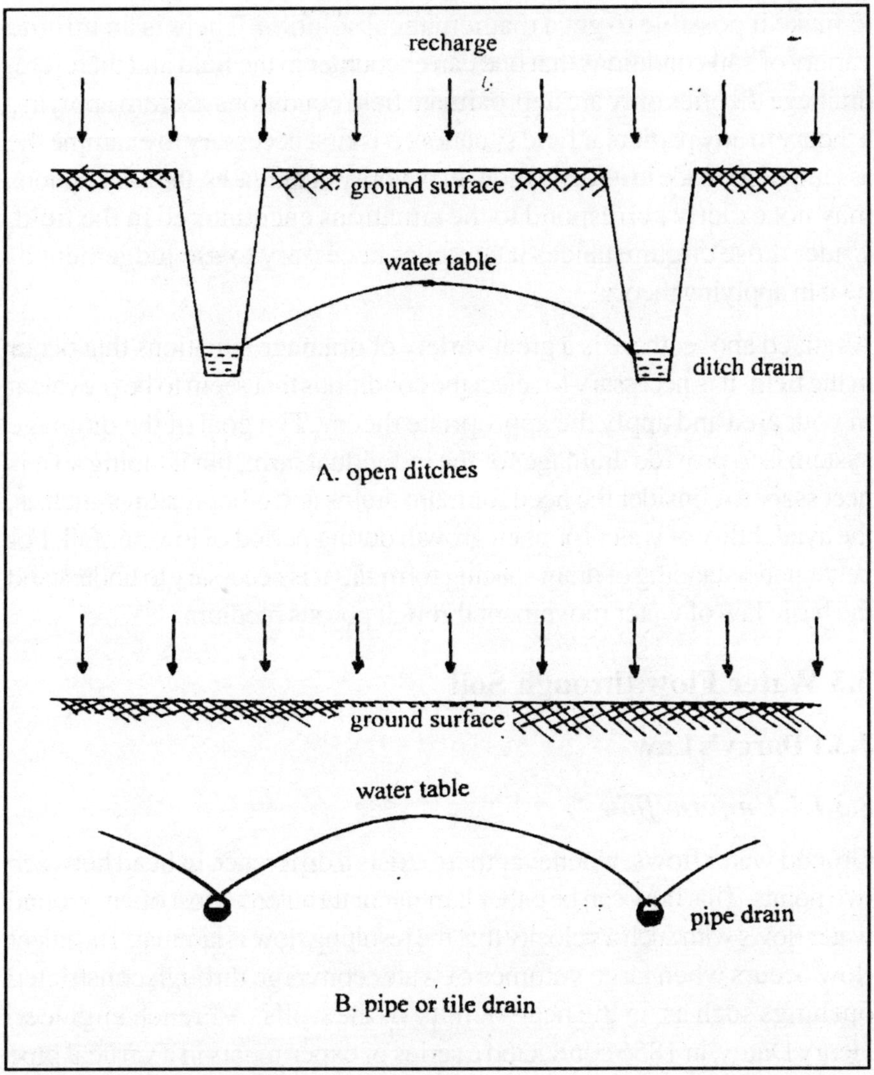

Figure 3.2. Cross section of drain laterals

- A uniform distribution of recharges, steady or time dependent, over the area between the parallel drains.
- Most of the drainage equations are based on Dupuit-Forchheimer assumptions.

A drainage theory describes the flow of water in an idealized soil-water system. Actual field problem is simplified through acceptable assumptions to make it possible to get a mathematical solution. There is an infinite variety of soil conditions that one can encounter in the field and therefore, drainage theories only are approximate field conditions. Before applying a theory to any particular field situation, it is first necessary to examine the assumptions made in its derivation. In majority of the cases, the assumptions may not exactly correspond to the situations encountered in the field. Under those circumstances it becomes necessary to use judgement of own in applying theory.

As stated above, there is a great variety of drainage situations that occur in the field. It is necessary to select the conditions that seem to be prevalent in your area and apply the appropriate theory. The goal of the drainage system is to provide drainage for the individual farm, but in doing so it is necessary to consider the need for main drains and other features such as, the availability of water for plant growth during period of low rainfall. For better understanding of drain spacing formulas it is necessary to understand the basic law of water movement through porous medium.

3.3 Water Flow through Soil

3.3.1 Darcy's Law

3.3.1.1 Uniform flow

Ground water flows, whenever there exists a difference in head between two points. This flow can be either laminar or turbulent. Most often, ground water flows with such a velocity that the resulting flow is laminar. Turbulent flow occurs when large volumes of water converge through constricted openings such as, in the near vicinity of the wells. A French engineer, Henry Darcy, in 1856 conducted a series of experiments in a vertical pipe filled with sand to understand the flow behavior of water in porous medium. He concluded that the rate of flow through a column of saturated sand is proportional to the difference in hydraulic head at the end of column and also to the area of flow cross-section and inversely proportional to the length of the column. **Figure 3.3** illustrates the Darcy's experiment in which there is a saturated ground water flow induced by a difference of water levels. Darcy's law can be mathematically expressed as below :

$$Q = K.A.h_n / L \qquad 3.1$$

where,

$Q =$ discharge or flow $[L^3/T]$

$K =$ saturated hydraulic conductivity or permeability of the soil $[L/T]$

$A =$ cross sectional area of the soil through which flow takes place $[L^2]$

$h_n =$ potential loss or hydraulic head loss or piezometric pressure loss over a length of L [L]

$L =$ distance/length in the direction of flow over which h_n takes place [L]

The ratio of h_n/L is also called as average potential gradient, hydraulic gradient or piezometric gradient. In the experiment, the soil is assumed to have uniform hydraulic conductivity (K). If cross section (A) in the direction of flow is constant, as in the experiment then the hydraulic gradient will also be constant at any place and equal to h_n/L. In the experiment the flow occurs between two confining layers and the discharge (Q) through any cross sectional area is the same. This is known as confined unvaried flow.

When A is rectangular having height (d) and width (B), so that A =B .d, then Eqn. (3.1) can be written as below.

$$\frac{Q}{B} = Q' = K.d.\frac{h_n}{L} \qquad 3.2$$

where,

$Q' =$ discharge per unit width $[L^3/L/T]$

The product Kd is called as transmissivity. Considering $v = Q/A = Q'/d$ Eqn. (3.1) can be also expressed as,

$$v = K.\frac{h_n}{L} \qquad 3.3$$

where,

v = apparent flow velocity or specific discharge or flux $[L^3/T/L^2$ or $L/T]$

When v and h_n/L are constant at any place, the **flow is called as uniform flow.** However the actual flow velocity in the soil mass is higher than v because the water passes through pores having total cross sectional area smaller than A.

Equation. (3.3) can also be written in the following form

$$h_n = v.\frac{L}{K}$$

3.4

From analogy of Ohm's law for electric flow, the ratio L/K can be called resistance to uniform flow. Equation 3.1 can also be generalized as,

$$v = K\frac{(\phi_1 - \phi_2)}{(s_2 - s_1)} = -K\frac{\Delta\phi}{\Delta s}$$

3.5

where,

$\Delta\phi = h_n = \phi_1 - \phi_2$

$\Delta s = s_1 - s_2$

ϕ_1 and ϕ_2 are potential at plane 1 and 2, respectively, and

s_1 and s_2 are distance of plane 1 and plane 2

3.3.1.2 Non-uniform flow

When the cross sectional area changes in the direction of flow, the velocity must be changing in the direction of flow because the flow is unvaried (**Figure 3.4**). Similarly the hydraulic gradient will also change in the direction of flow. When the cross sectional area is small, the velocity and hydraulic gradient are large and vice-versa. Darcy law for non-uniform flow can be written as follow.

$$v_s = -K_s.\frac{\delta\phi}{\delta s}$$

3.6

where,

s = direction of flow

v_s = flux at a certain point (x, y, z) in s-direction [L/T]

K_s = hydraulic conductivity at the same point in s-direction [L/T]

ϕ = potential at the same point [L]

$\delta\phi$ = change in potential over δs [L]

δs = small step form same point in s-direction [L]

The ratio of $\delta\phi/\delta s$ is called the local potential gradient, hydraulic gradient or piezometric gradient in s-direction. It is also called as differential coefficient of ϕ in s-direction.

Since water flows from a higher elevation (high pressure) to low elevation (low pressure), the potential ϕ can also be written as sum of elevation head, z and the pressure head, p as follows.

$$\phi = p + z \qquad 3.7$$

Now the Darcy law can be expressed as below.

$$v_s = -K_s.(\frac{\delta\phi}{\delta s} + \frac{\delta z}{\delta s}) \qquad 3.8$$

3.3.1.3 Water table and equipotential

Any line or plane, straight or curved, through which water is passing and in which ϕ is constant is called equipotential. Lines or plane perpendicular to equipotential lines are called stream lines or flow lines.

The pressure along a water table is atmospheric which is often taken as reference pressure and set to be zero. Above water table pressure is negative and below water table pressure is positive. The pressure gradient $\delta\phi/\delta w$ in the direction w of water table is zero. Hence the flow velocity in the direction of water table equals.

$$v_w = -K_w \frac{\delta H}{\delta w} \qquad 3.9$$

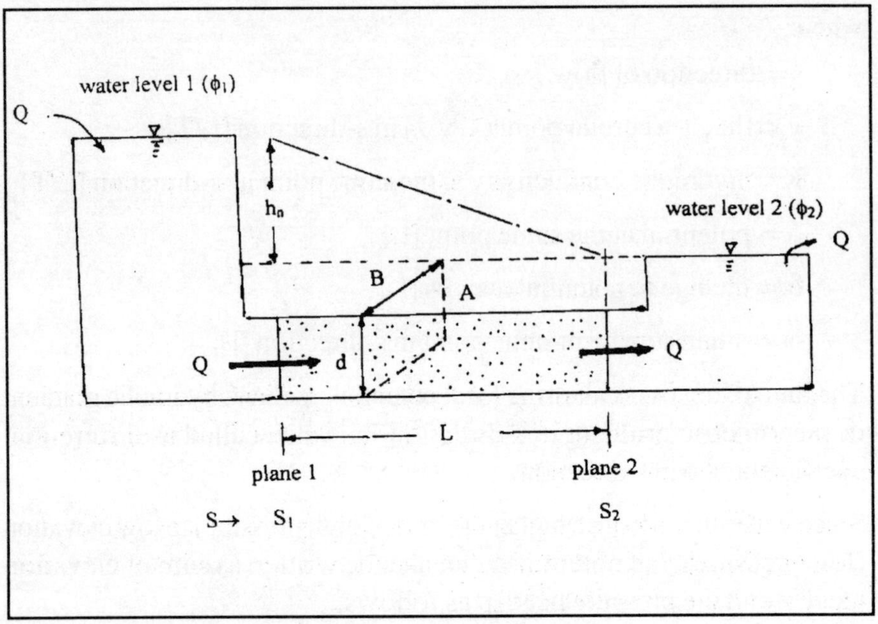

Figure 3.3. Darey's experiment for confined uniform flow.

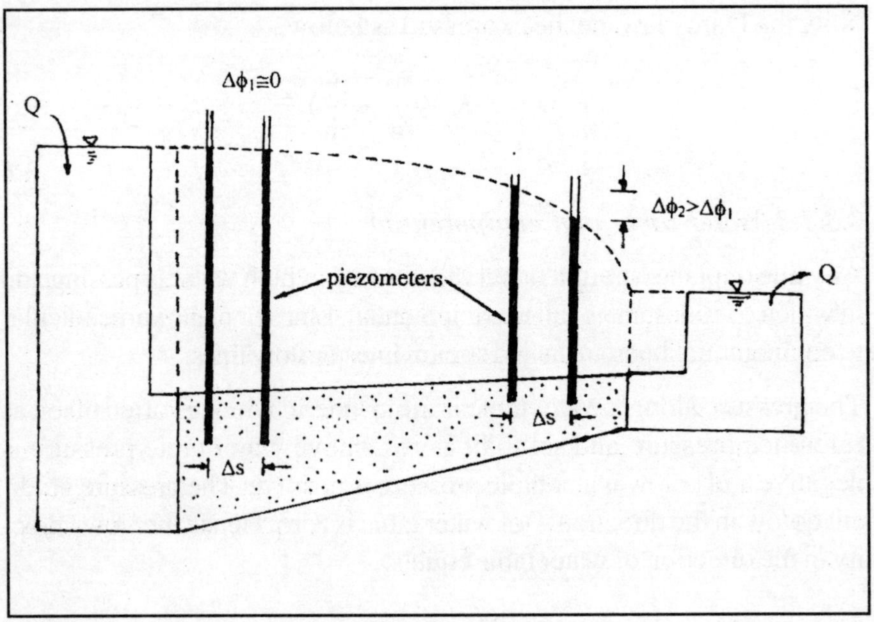

Figure 3.4. Confined non-uniform flow.

Equation (3.9) follows from Equation (3.8) using w instead of s and H instead of z and $\delta p/\delta w=0$. It is not valid for direction other than w. This means potential at the water table cannot be taken equal to the height of water table: $\phi \neq H$, unless the water table is horizontal. For horizontal direction x, one can find for example that $v_x \neq -K_x\, \delta H/\delta x$, because $\delta p/\delta x \neq 0$ unless water table is horizontal (but then $v_x = 0$). In vertical direction z one find $v_z \neq K_z\, \delta H/\delta z$ because $\delta H/\delta z=0$ and v_z need not to be zero.

To describe the flow in the presence of free water table, following simplifying assumptions are mostly used.

a) The water table is a stream line.

b) Near the water table the flow is only vertical.

c) Near the water table $\delta\phi/\delta z=0$ and $\delta\phi/\delta x=\delta H/\delta x$ (Dupuit assumptions)

Above assumptions will be used frequently in coming sections.

3.3.2 Flow Through Parallel Soil Layers

Figure 3.5 shows the one-dimensional flow through parallel soil layers, with the flow parallel to the interfaces of the layers. Each layer has different hydraulic conductivity (K_1, K_2, K_3,...., K_n) and thickness (d_1, d_2, d_3, ...d_n). Flow velocity in different soil layers is different but the hydraulic gradient is constant at any layer. Total flow can be expressed by an equivalent uniform flow with equivalent transmissivity, Te. Using Equation (3.2) the total flow through all layers Q′t can then be written as,

$$Q't = Te\frac{h_n}{L} \tag{3.10}$$

where,

$Te = K_e D_t$ [L^2/T]

K_e = equivalent hydraulic conductivity of all layers,[L/T]

$D_t = d_1 + d_2 + d_3 + \ldots d_n$ [L]

For each layer one can write the discharge as,

$Q_1{}' = K_1\, d_1\, h_n/L$, $Q_2{}' = K_2\, d_2\, h_n/L$, $Q_3{}' = K_3\, d_3\, h_n/L$, etc.

Since $Q'_t = Q'_1 + Q'_2 + Q'_3 + \dots$

$\therefore\ Q'_t = (K_1 d_1 + K_2 d_2 + K_3 d_3 + \dots K_n d_n)\, h_n/L = (\Sigma K.d)\, h_n/L$

$$Te = K_1 d_1 + K_2 d_2 + K_3 d_3 + ... + K_n d_n = \sum Kd \qquad 3.11$$

3.3.3 Flow Perpendicular to Soil Layers

Figure 3.6 shows the one dimensional non-uniform flow through parallel soil layers extending perpendicular to the direction of flow. Each layer has different hydraulic conductivity (K_1, K_2, K_3, ...K_n) and thickness (d_1, d_2, d_3,....d_n). Piezometric levels at the interfaces between the layers are shown in figure. h_1, h_2, and h_3 are the head losses in each layer. Velocity of flow is same for all the layers but the hydraulic gradients are different (K_1/d_1, K_2/d_2 K_3/d_3, ...K_n/d_n). From the Equation (3.4) total head loss can be calculated as,

$$h_t = v.\frac{D_t}{K_e} \qquad 3.12$$

where,

$D_t = d_1 + d_2, + d_3, \dots + d_{n,}$, [L]

K_e = equivalent hydraulic conductivity of all layers, [L/T]

For each layer one may also write.

$h_1 = v\, d_1/K_1$, $h_2 = v d_2/K_2$, $h_3 = v d_3/K_3$, ...$h_n = v d_n/K_n$

Since $h_t = h_1 + h_2, + h_3, \dots + h_n$

$D_t/h_t = (d_1/K_1 + d_2/K_2 + d_3/K_3 \dots + d_n/K_n) v$

$D_t/K_e = (d_1/K_1 + d_2/K_2 + d_3/K_3 \dots + d_n/K_n) = \Sigma d/K$

Which shows that the resistance is additive. The resistance to one

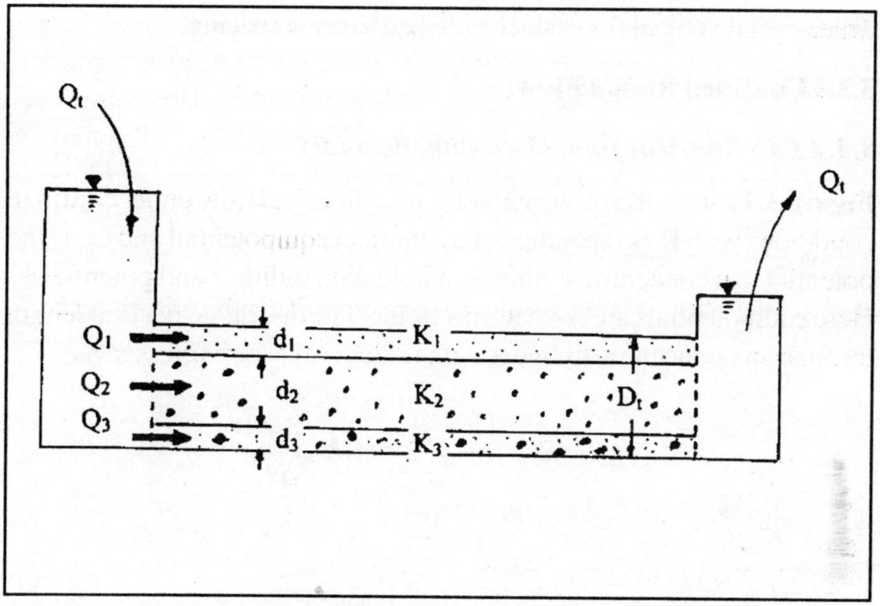

Figure 3.5. Flow parallel to soil layers.

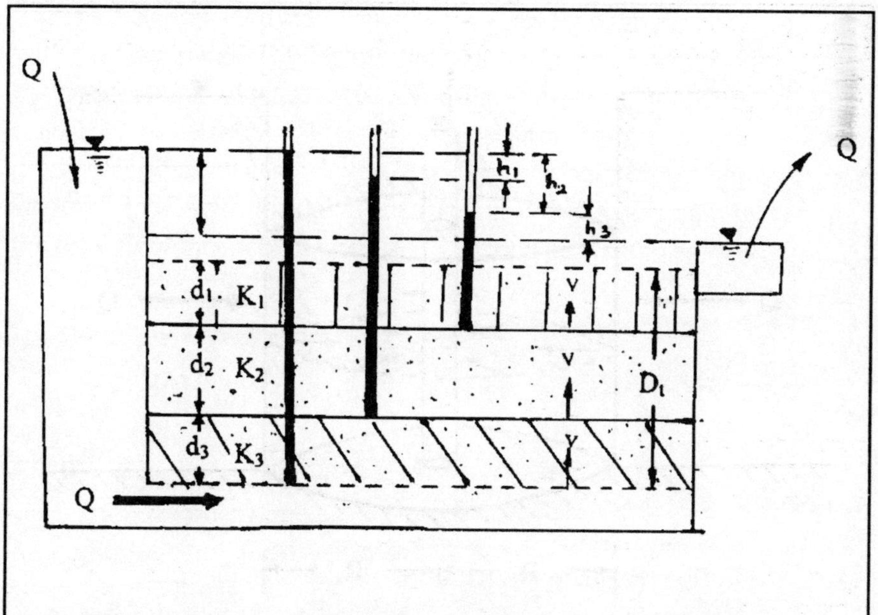

Figure 3.6. Flow perpendicular to soil layers.

dimensional vertical flow may be applied to drain spacing.

3.3.4 Confined Radial Flow

3.3.4.1 Cylindrical flow (Two dimensional)

Figure 3.7 shows the flow geometry of cylindrical flow under confined conditions with R as the radius of cylindrical equipotential and Q_R is the potential to a concentric cylindrical hole with radius, r and potential ϕ_r. Here equipotentials are concentric circles. The discharge per unit length through any concentric cylinder with radius x may be expressed as,

$$Q' = 2\pi .x.v = 2\pi .x.K.\frac{\delta\phi}{\delta x}$$

$$\delta\phi = \frac{Q'}{2\pi K}.\frac{\delta x}{x}$$

Integrating the above equation for limits x = r, $\phi=\phi_r$ and x = R, $\phi=\phi_R$, one

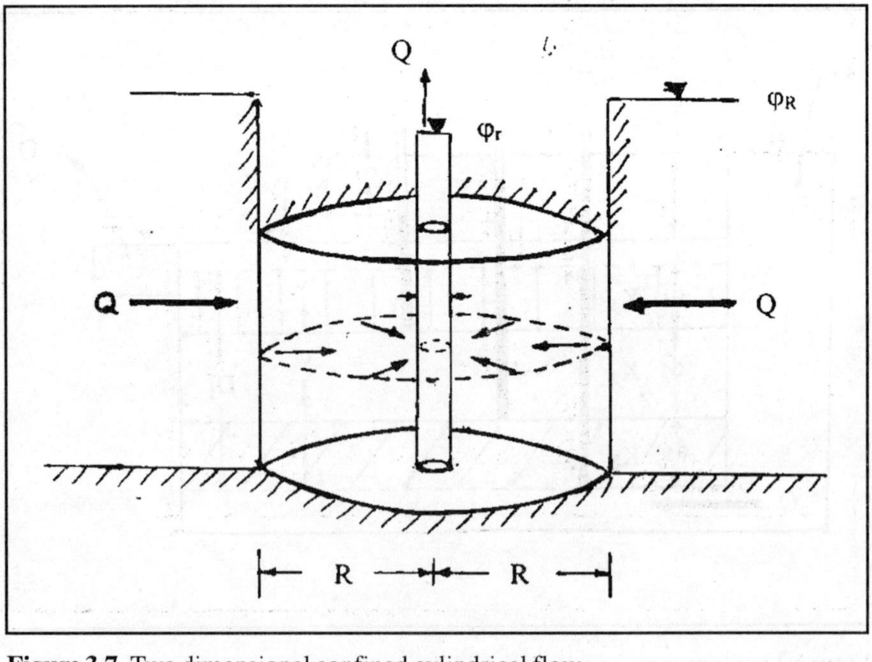

Figure 3.7. Two dimensional confined cylindrical flow.

may arrive at,

$$\phi_R - \phi_r = \frac{Q'}{2\pi K} \ln \frac{R}{r}$$

Here (ϕ-ϕ_R) is head loss and can be written as h_n. Hence the above equation becomes.

$$Q' \ln \frac{R}{r} = 2\pi K h_n$$

3.13

If the hole is horizontal then the flow will take place in half cylinder only. Therefore, Q' reduces to half of it's value and Equation (3.13) reduces to:

$$Q' \ln \frac{R}{r} = \pi K h_n$$

3.14

3.3.4.2 Spherical flow (Three dimensional flow)

Figure 3.8 shows the flow from a spherical equipotential with radius R and potential ϕ_R.

The discharge through any concentric sphere with radius x can be expressed as,

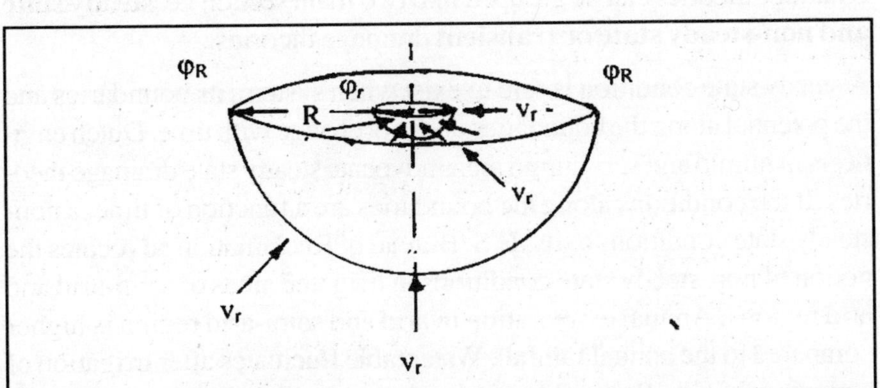

Figure 3.8. Three dimensional spherical flow.

$$Q = 4\pi x^2.v = 4\pi x^2.K\frac{\delta\phi}{\delta x}$$

$$or, \delta\phi = \frac{Q}{4\pi K}\frac{\delta x}{x^2}$$

Integrating the above equation under the limits $x = r, \phi = \phi_{r,}$ $x = R, \phi = \phi_{R,}$ one get.

$$\phi_R - \phi_r = h_n = \frac{Q}{4\pi K}(\frac{1}{r} - \frac{1}{R})$$

3.15

Since 1/R<<<1/r, above equation may be also written as,

$$Q = 4\pi K r h_n$$

3.16

If flow occurs only in half of the sphere, the discharge Q will reduce to half of its value.

$$Q = 2\pi K r h_n$$

3.17

3.4 Steady State Drainage Theories

Drainage theories can be grouped into two main section i.e. **steady state and non-steady state or transient** drainage theories.

A steady state condition is said to exist when system, its boundaries and the potential along the boundaries does not change with time. Dutch engineers in humid and sub-humid areas advocate steady state drainage theories. If the conditions along the boundaries are a function of time, a non-steady state conditions exits. U.S. Bureau of Reclamation advocates the notion of non-steady state conditions in irrigated areas of semi-arid and arid regions. Annual evaporation in arid and semi-arid region is higher compared to the annual rainfall. Water table fluctuates after irrigation of agricultural fields. Salt build up is a common problem in irrigated agriculture. The drains are designed to lower water table positions in these areas to facilitate leaching of excessive salts present in the root zone and to check salt build up due to capillary rise. Salt build up is not a problem in humid region as the annual rainfall is higher than the annual evaporation and down ward movement of salt and flushing of root zone is high and frequent. The basic aim of pipe drains in sub-humid and humid region is to control only water table positions. Thus the depths of pipe drains below the ground surface is more in irrigated areas of arid and semi arid regions.

3.4.1 The Dupuit-Forchheimer Assumptions

It is a difficult task to describe ground water flow with a phreatic surface. For better approximation of flow of ground water some simplifying assumptions are used commonly known as Dupuit-Forchheimer assumptions, written as below.

a) The flow meets resistance only in horizontal direction.

b) Horizontal flow velocity is constant within each vertical cross section.

c) Within each vertical cross section the horizontal hydraulic gradient is constant and, equal to the slope of water table.

These assumptions do not cause considerable error when horizontal component of flow paths is much longer than the vertical component or when vertical hydraulic conductivity is not much less than horizontal one.

3.4.2 Derivation of Ellipse Equation

One of the earliest efforts to find a mathematical solution to a drainage problem resulted in the development of an elliptic equation for the shape of water table over the tile drains. **Colding (1872)**, a Danish engineer first presented this equation. **Rothe (1924)**, **Kozeny (1932)** and **Hooghoudt (1937)** developed it independently. **Russel (1934)** of Rothe's work finds the first mention in English in a review. **Aronovici and Donnan (1946)**, not aware of the European work, gave the first derivation of the equation in American literature. Hooghoudt's reasoning is followed here. **Figure 3.9** represents a homogeneous soil underlain by an impermeable layer and drained by parallel vertical walled open ditches which penetrates to the impermeable layer. Assuming that a constant rate of rainfall is removed equally well at all distances from the drain, the rate q_x at which

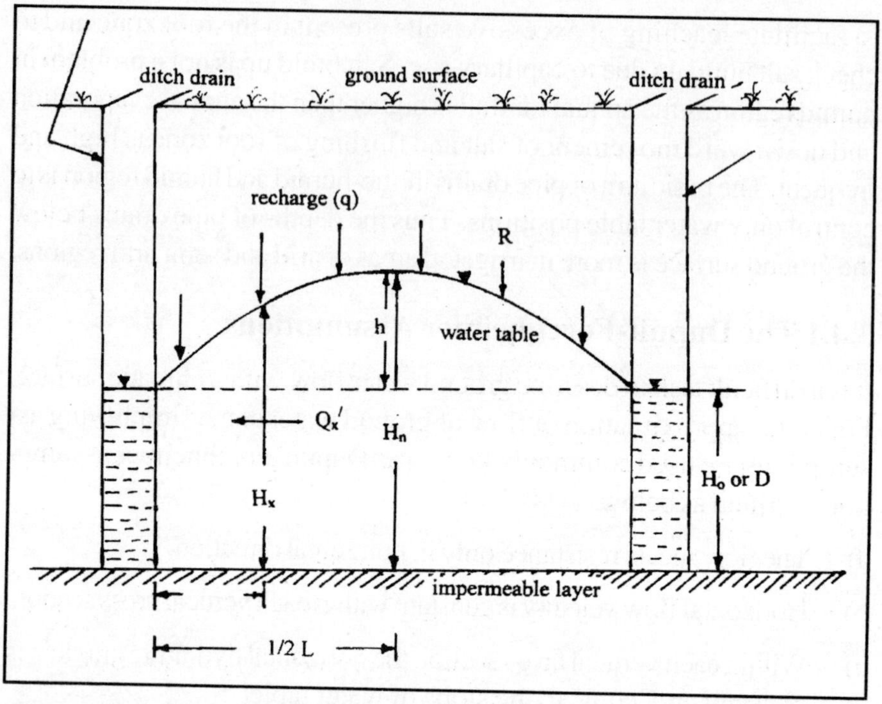

Figure 3.9. Parallel open resting on impervious layer

water crosses a vertical plane at any x can be written as.

$$Q = q\frac{L/2 - x}{L/2}$$

Where,

q_x= rate at which water crosses a vertical plane at any distance, x [L/T]

S= L/2 = half drain spacing [L]

L = drain spacing, [L]

x = any distance x in horizontal direction [L]

Q = q/2, one half the total discharge of each drain per unit length [L]

q = drain discharge rate per unit drained area [L/T]

From Dupuit assumption it follows that

$$q_x = -yv_x = yK\frac{dy}{dx}$$

Equating the two expressions for q_x, one will obtain the following differential equation.

$$ydy = (2Q/LK)(L/2 - x)dx$$

Integrating for the limit, at x=0, y=Ho and at x=x , y=H_x, will yield.

$$H_x^2 - Ho^2 = (2Q/LK)(Lx - x^2)$$

where,

H_o = height of water level in the open ditch above impervious layer, [L]

This is the equation of ellipse. Above equation can be further simplified by substituting x=L/2 and y=H_n for the midpoint between the drains. Above equation can be written as.

$$L = 2K(H^2{}_n - H_0{}^2)/Q$$

$$L = 4K(H_n{}^2 - H_o{}^2)/q$$

In this equation if other quantities are known then drain spacing can be calculated.

Colding analysis differed from the above analysis as he assumed H_0 to be zero. From experimental observation he determined that the spacing between the drains is about 0.9 times the major axis of the ellipse equation. Colding's final result can be written as,

$$L = 1.8H\sqrt{\frac{K}{R}}$$

where,

R = rainfall rate (=2Q /L)

3.4.3 Derivation of Hooghoudt Equation

3.4.3.1 First Approach

Figure 3.9 shows two parallel open drains (ditches) penetrating to horizontal impermeable layer. Various assumptions involved in the derivation of Hooghoudt's steady state drain spacing formula are written below.

1. Soil is homogeneous.
2. The drains are evenly placed at a distance of L apart.
3. Dupuit-Forchheimer assumptions hold good.
4. Darcy's law is valid.
5. Impermeable layer underlies the drain at a depth D.
6. Rainfall, recharge or irrigation water is applied at a rate of q.
7. Origin of coordinates is taken on the impermeable layer below the centre of the one drain.

Assumption 3 implies that water flows horizontally because all the equipotentials are vertical planes. This is an erroneous picture of actual flow path of water. This is specially incorrect near the drains where the flow paths are quite curved. However, where the slope of water table is relatively flat the Dupuit-Forchheimer assumptions are nearly valid. Assumptions cause only 10% inaccuracy in the result.

The discharge per meter length of drain through a cross section at a distance x from the drain equals.

$$Q'_x = q(\frac{1}{2}L - x) \qquad 3.18$$

Using the Darcy's law and the Dupuit-Forchheimer assumptions the flow passing through the cross section per unit distance perpendicular to the plane of the drawing.

$$Q'_x = H_x v = -KH_x \frac{\delta H_x}{\delta x}$$
$$= -\frac{K}{2}\cdot\frac{\delta}{\delta x}(H_x)^2$$

3.19

where,

H_x = height of water table above impervious layer at a section x distance apart, [L]

From Equation (3.18) and (3.19)

$$q(\frac{1}{2}L - x)\delta x = \frac{1}{2}K\delta.H_x{}^2$$

Integrating the above differential equation under the conditions that $H_x = H_o$ at x = 0 and $H_x = H_n$ at x=1/2 L

$$\int_0^{L/2} q(L\delta x - \delta x^2) = \int_{H0}^{Hn} K.\delta H_x{}^2$$

$$q(\frac{1}{2}L^2 - \frac{1}{4}L^2) = K(H_n{}^2 - H_0{}^2) = K(H_n + H_0)(H_n - H_0)$$

Hence H_n-H_o = h and $H_n + H_o = H_o + h + H_o = 2H_o + h$

Substituting above values in above solution will give,

$$\frac{1}{4}qL^2 = K(2H_0 + h)h = 2KH_0h + Kh^2$$
$$qL^2 = 8KH_0h + 4Kh^2$$

3.20

Top level of water in ditch (H_0) is drainage level, therefore H_0 can be set equal to D which is the depth of impervious layer below drainage level.

Equation (3.20) can be written as below.

$$L^2 = \frac{8KDh + 4Kh^2}{q} \quad 3.21$$

This is Hooghoudt drain spacing formula in homogeneous soil when drainage is well above the impervious layer.

If water level in drainage ditch or drain is at zero level i.e. $D = H_o = 0$, then the Equation (3.21) reduces to the following form :

$$q = \frac{4Kh^2}{L^2} \quad 3.22$$

This represents the flow above the drain level consequently the term $8KDh/L^2$ corresponds to the flow below drain level. One can see that from Equation (3.21)

$$q_a = \frac{4Kh^2}{L^2} \quad 3.23$$

stands for the horizontal flow above drain level, and

$$q_b = \frac{8KDh}{L^2} \quad 3.24$$

stands for the horizontal flow below the drain level. The total flow, thus, becomes,

$$q = q_a + q_b \quad 3.25$$

3.4.3.2 Second Approach

The definition sketch of the flow situation is shown in **Figure 3.10** which shows a homogeneous soil of known hydraulic conductivity K. Impermeable layer lying at a distance D from the bottom of the ditch or pipe/drain. The soil is drained by a series of parallel ditches.

The discharge per meter length of drain through a cross section at a distance x from the origin can be written as.

$$q_x = (\frac{L}{2} - x)q \qquad 3.26$$

A second equation can be obtained for q_x by applying Darcy's law to flow through plane. Using the assumption that the hydraulic gradient is equal to the slope of water table at any point, one may easily write.

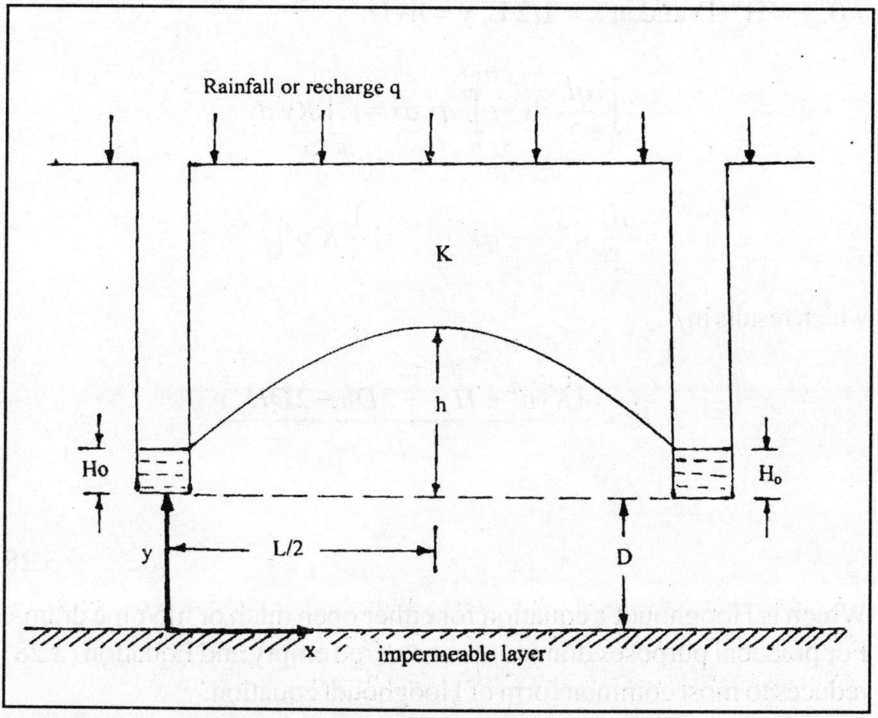

Figure 3.10. Definition sketch of Hooghoudt drain spacing formula.

$$q_x = Ky\frac{dy}{dx}$$

3.27

Right hand side of Equation (3.26) and Equation (3.27) must be equal since the flow in the two cases are the same. Therefore,

$$(\frac{L}{2} - x)q = Ky\frac{dy}{dx}$$

$$(\frac{L}{2} - x)q.dx = Ky.dy$$

$$\frac{qL}{2}dx - qx.dx = Ky.dy$$

This ordinary differential equation can be solved under the limits, when x = 0, y = H_o+D and at x = 1/2 L, y = h+D.

$$\int_0^{L/2}\frac{qL}{2}dx - \int_0^{L/2} qx.dx = \int_{Ho+D}^{h+D} Ky.dy$$

$$\frac{qL}{2}x\Big|_0^{L/2} - qx^2\Big|_0^{L/2} = \frac{1}{2}Ky^2\Big|_0^{L/2}$$

which results in

$$L^2 = \frac{4K(h^2 - H_o^{\ 2} + 2Dh - 2DH_o)}{q}$$

3.28

Which is Hooghoudt's equation for either open ditch or tile/pipe drains. For practical purposes drains are considered empty and Equation (3.28) reduces to most common form of Hooghoudt equation.

$$L^2 = \frac{8KDh + 4Kh^2}{q}$$

when Ho=0 3.29

3.4.3.3 Third Approach

Hooghoudt Equation (3.28) can also be derived with the help of Boussinesq equation of ground water flow, which is written below.

$$\frac{\partial}{\partial x}(Ky\frac{\partial y}{\partial x}) + p = f\frac{dy}{dt}$$

3.30

Above equation can be changed to steady state conditions as below.

$$\frac{\partial}{\partial x}(Ky\frac{\partial y}{\partial x}) + p = 0$$

3.31

This differential equation can be solved as follows.

$$\frac{\partial}{\partial x}(Ky\frac{\partial y}{\partial x}) = -p$$

$$Ky\frac{\partial y}{\partial x} = -px + C_1$$

$$y\frac{\partial y}{\partial x} = -\frac{px}{K} + \frac{C_1}{K}$$

$$y^2 = -\frac{px^2}{K} + C_1x + C_2$$

3.32

For given boundary conditions

x = 0, y = Ho + D (1)

x = 1/2 L, y = h + D (2)

x = L, y = Ho + D (3)

Constant can be worked out as follows using above boundary conditions.

Substituting boundary condition (1) in Equation (3.32)

$$(Ho+D)^2 = 0 + 0 + C_2$$

$$y = -p\frac{x^2}{K} + C_1x + (H_o + D)^2$$

Putting boundary conditions (3) into above equation will yield,

$$(H_o + D)^2 = -\frac{pL^2}{K} + C_1L + (H_o + D)^2$$

$$C_1 = \frac{pL}{K}$$

Therefore Equation (3.32) can be finally written as,

$$y^2 = -\frac{px^2}{K} + \frac{pL}{K}x + (H_0 + D)^2$$

From boundary condition (2)

$$(h + D)^2 = -\frac{pL^2}{4K} + \frac{pL^2}{2K} + (H_0 + D)^2$$

When drains run empty the above equation reduces to the following form:

$$L^2 = \frac{4K}{p}(2Dh + h^2)$$

where,

p= steady replenishment equal to drain discharge, q.

Therefore, above equation can be rewritten as,

$$L^2 = \frac{8KDh + 4Kh^2}{q}$$

3.33

Which is well known Hooghoudt equation.

3.4.4 The maximum depth of flow

In Hooghoudt steady state drain spacing equation derived in earlier sections the resistance to vertical flow has been ignored. Assuming that the soil has a hydraulic conductivity for vertical flow K_v, then the hydraulic head required to overcome the resistance to vertical flow h_v becomes in accordance to Equation (3.3) (replacing v by q and L by $H_n = H_o + h$).

$$h_v = \frac{q(H_o + h)}{Kv}$$

3.34

The hydraulic head required to overcome the horizontal resistance, following Equation (3.20) is,

$$h_h = \frac{qL^2}{8K_h H_o + 4K_h h}$$

3.35

where,

K_h = horizontal hydraulic conductivity, [L/T]

Setting the total head $h = h_v + h_h$, one may get from Equation (3.34) and (3.35).

$$\frac{h}{q} = \frac{L^2}{8K_h.H_o + 4K_h.h} + \frac{H_o + h}{K_v}$$

Minimizing h/q as a function of Ho and assuming that the term $4K_h$ h is negligibly small as compared to 8 K_h Ho and the term h is also very small compared to H_o, it is found that.

$$\frac{h}{q} = \frac{L^2}{8K_h H_o} + \frac{H_o}{K_v} \tag{3.36}$$

Differentiating the above equation with respect to H_o one will get the following equation.

$$\frac{1}{q}\frac{dh}{dH_o} = -\frac{L^2}{8K_h H_o^{\ 2}} + \frac{1}{K_v} = 0 \tag{3.37}$$

So that maximum value of H_o, expressed by H_m

$$\frac{L^2}{H_m^{\ 2}.8K_h} = \frac{1}{K_v}$$

$$\frac{L}{H_m} = \sqrt{\frac{8K_h}{K_v}} \tag{3.38}$$

If soil is isotropic ($K_h = K_v$) one obtain

$$H_m = \frac{L}{\sqrt{8}} = 0.35L \tag{3.39}$$

If soil is anisotropic with $K_v < K_h$ the ratio Ho/L is less than 0.35 but when $K_v > K_h$ then H_o/L can be greater than 0.35.

Hence it can be concluded that the depth H_o (=D) cannot be greater than H_m=0.35 L

3.4.5 Transformation of Hooghoudt equation into ellipse equation

The Hooghoudt equation can be changed to the form of ellipse equation by transforming the origin of the coordinate system to the midpoint between the drains. Considering the governing differential equation for the derivation of Hooghoudt equation, which is written below.

$$q(L/2)\,dx - q.x.dx = K.y.dy$$

After integration, following general solution of the phreatic surface may be obtained.

$$\frac{qL}{2}x - \frac{qx^2}{2} = K\frac{y^2}{2}$$

3.40

To move the origin of coordinate system to the midpoint between the drains, one should substitute the transformation, $x_1 = L/2 - x$. On substitution one will obtain.

$$\frac{y^2}{\frac{L^2 q}{4K}} + \frac{x^2}{\frac{L^2}{4}} = 1$$

3.41

Which is the equation of ellipse having semi-major and semi-minor axes given by L/2 and $L/2(q/K)^{1/2}$, respectively.

3.4.6 Hooghoudt Equations for Different Drainage Conditions

3.4.6.1 Pipe drains or ditches in homogeneous soil (*Ho<0.35 L*)

Figure 3.11 shows pipe drains and ditches in the situation as described in section

3.4.3.1. According to Equation (3.21) the spacing equation becomes.

$$L = \sqrt{\frac{8KDh + 4Kh^2}{q}} \qquad 3.42$$

For pipe drains which are just filled, D=2r (drain diameter) so that

$$L = \sqrt{\frac{16K.r.h + 4Kh^2}{q}} \qquad 3.43$$

When the bottom of the ditches or pipe drains is at a short distance above the impermeable layer, the above formulae remain applicable. When the drainage level coincides with the impermeable layer both the above equation reduces to.

$$L = 2h\sqrt{\frac{K}{q}} \qquad 3.44$$

3.4.6.2. Pipe drains in two different soil layers with interface at drainage level when $K_a > K_b$

Denoting the hydraulic conductivity of upper layer, above the drainage level by K_a and the one of the lower layer below drainage level by K_b and considering the Equations (3.23) and (3.24) the spacing Equations (3.42) and (3.44) for ditches and pipe drains can be changed into.

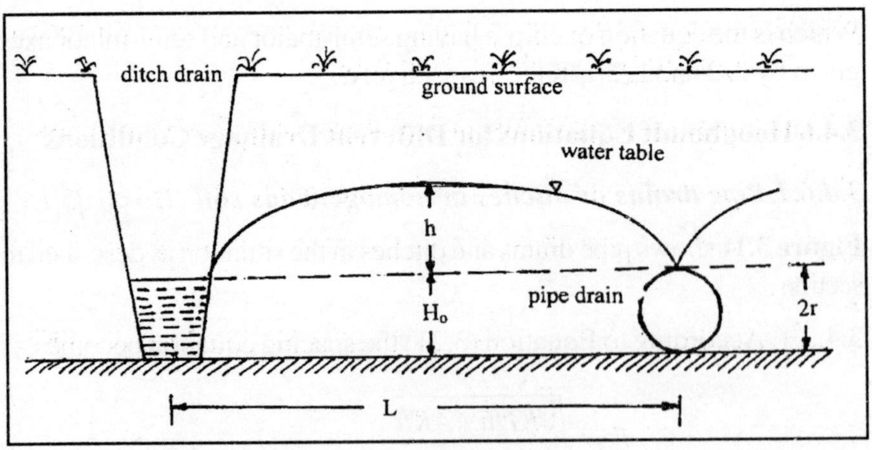

Figure 3.11. Pipe and ditch drains on impervious layer.

$$L = \sqrt{\frac{8K_b Dh + 4K_a h^2}{q}}$$

3.45

$$L = \sqrt{\frac{16K_b Dh + 4K_a h^2}{q}}$$

3.46

$$L = 2h\sqrt{\frac{K_a}{q}}$$

3.47

3.4.6.3 *Pipe drains in two different soil layers with interface at drainage level when $K_b > K_a$*

When the upper layer is relatively less permeable, the horizontal flow through this layer becomes very small. Recharging water is likely to pass vertically downward through the upper layer to proceed as horizontal flow only through the lower layer, more permeable layer when the hydraulic conductivity of upper layer (K_a) is very small the head loss due to vertical flow in upper layer become considerable. In Equation (3.3) ($v=K.h_n/L$). We can replace the v by q, K by K_a, h_n by h_v (hydraulic head required to overcome vertical resistance) and L by h (depth over which vertical flow takes place). We will find.

$$h_v = \frac{qh}{K_a}$$

3.48

Substituting the head loss h_v from the hydraulic head h, we obtain.

$$h = h_h + h_v = h - q\,(h/K_a)$$

$$h_h = h\,(1\text{-}q/K_a)$$

3.49

This is hydraulic head available for horizontal flow. Thus Equations (3.45) to (3.47) can be changed for such case by replacing h by h(1-q/Ka).

$$L=\sqrt{\frac{8K_bD.h(1-\frac{q}{K_a})+4K_a[h(1-\frac{q}{K_a})]^2}{q}} \qquad 3.50$$

$$L=\sqrt{\frac{16K_brh(1-\frac{q}{K_a})+4K_a[h(1-\frac{q}{K_a})]^2}{q}} \qquad 3.51$$

$$L=2h(1-\frac{q}{K_a})\sqrt{\frac{K_a}{q}} \qquad 3.52$$

From Equation (3.49) it can be seen that when $K_a<q$, there is no solution for L because the factor $(1\text{-}q/K_a)$ becomes zero or negative which physically mean that the vertical downward flow q cannot be more than K_a. If the rate of recharge (rainfall or irrigation) is greater than K_a then the excess will accumulate leading to the surface drainage problem.

3.4.7 Hooghoudt Drain Spacing Formulas when Drains Are Well Above Impermeable Layer

3.4.7.1 Uniform soil resistance to vertically cylindrical flow below drainage level dominating (D>0.1L)

Figure 3.12 represents the flow pattern when impermeable layer is at great depth. The hydraulic head required to overcome the resistance to the vertically cylindrical flow determines almost entirely the total flow. The horizontal flow above and below drainage level is negligible.

Hooghoudt (1940) derived the following equation for situation D>0.1L (where D is depth of impermeable layer below drainage level) using the method of images and superposition.

$$h = \frac{qL}{K} \ln \frac{L}{\pi r} \qquad 3.53$$

or

$$h = \frac{qK}{K} \ln \frac{L}{u} \qquad 3.54$$

where,

r = is the radius of the drain, [L]

u = π r, is the wetted perimeter of drain if half full, [L]

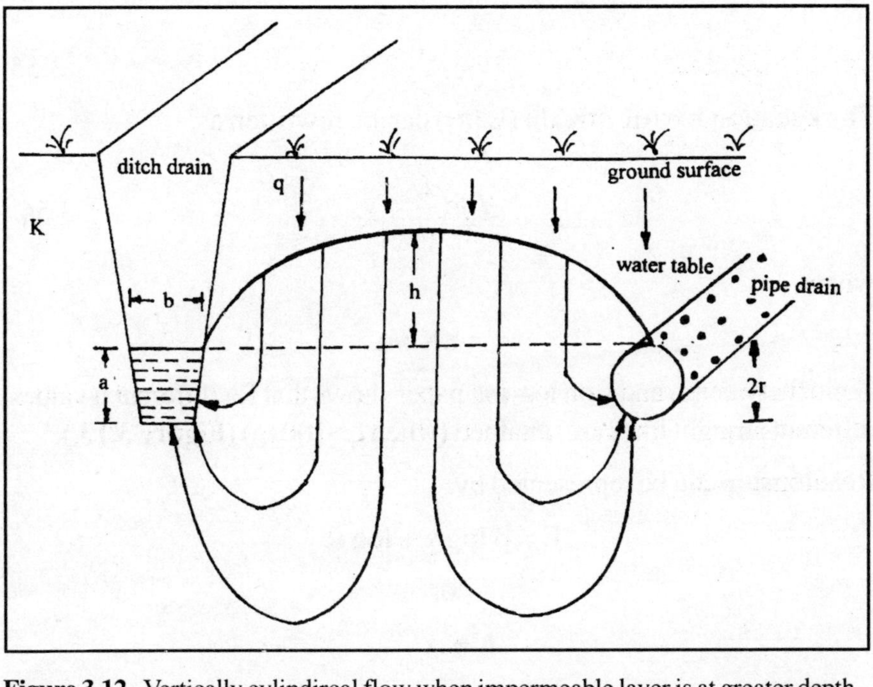

Figure 3.12. Vertically cylindircal flow when impermeable layer is at greater depth.

Equation (3.54) was also derived by some unknown authors before 1940 with the help of the theory of complex function. Equation (3.54) can also be derived by referring **Figure 3.3** and using the following differential equation.

$$d\phi = qL/2\pi K \ dx/x$$

$$Q'\ln R/r = \pi Kh, \quad Q' = qL$$

$$h = qL/\pi K \ln R/r$$

when cylinder is considered as horizontal then R can be replaced with L and r by πr. Finally we can write the above equation.

$h = qL/\pi K \ln L/\pi r$

For a ditch with height, a, of the water body and average width b of the water body the equivalent radius is

$$r = \frac{u}{\pi} = \frac{(a + 0.5b)}{2} \quad 3.55$$

The equation $h = (qL/\pi K) \ln (L/\pi r)$ can be rewritten as,

$$L \ln \frac{L}{u} = y \quad 3.56$$

where,

$$y = \pi Kh/q$$

A plot between L and y on log-log paper shows that for different u values different straight lines are obtained (when L>100 m) (**Figure 3.13.**).

Relationship can be represented by:

$$L = \beta \log y + \log \alpha$$

or

$$L = \alpha y^{\beta} \quad 3.57$$

The value of α and β can be obtained from **Figure 3.13** which bears a relation to log u as shown in **Figure 3.14**. Hence

$$\alpha = 0.56\ u^{0.29} \tag{3.58}$$

$$\beta = 0.85 - 0.046 \log u \tag{3.59}$$

Hooghoudt preferred to present his equations in terms of horizontal flow to imaginary ditches penetrating to an imaginary impermeable layer with a depth d below drainage level, which is termed as equivalent depth.

If horizontal flow above drainage level is neglected then Hooghoudt

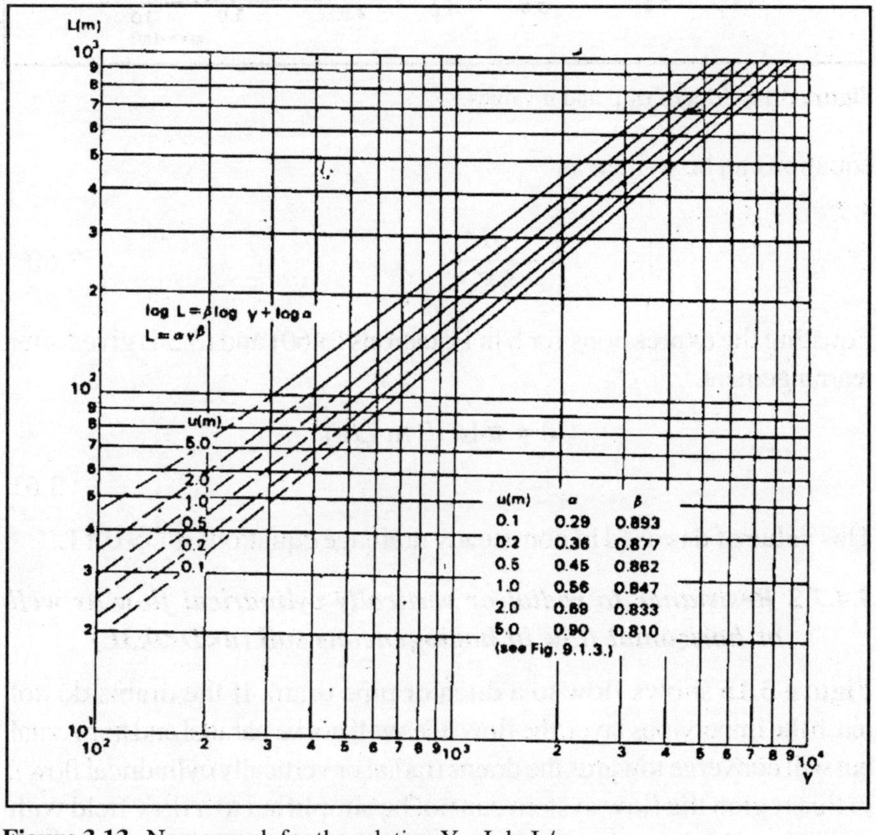

Figure 3.13. Nomograph for the relation Y = L ln L/u.

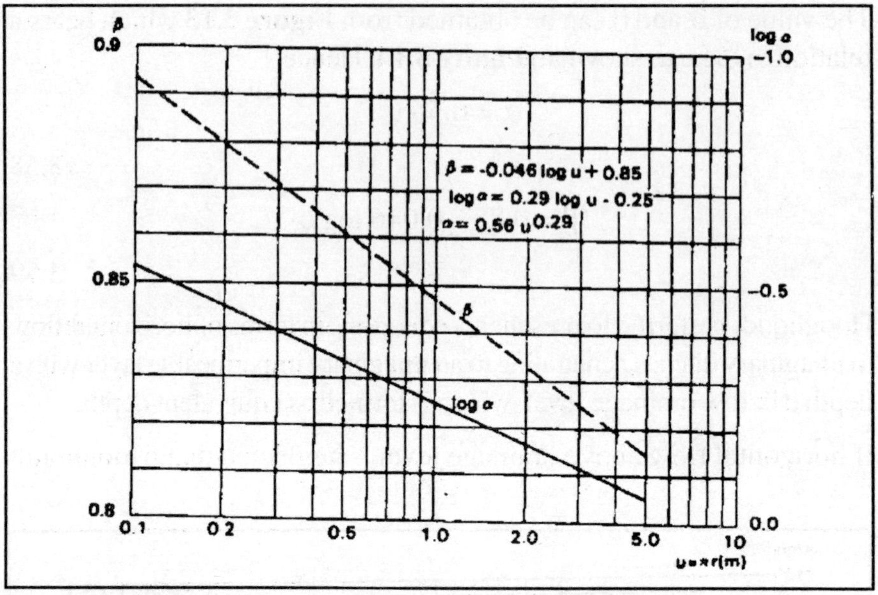

Figure 3.14. Plot of α, β and u values.

equation can be written as.

$$h = \frac{qL^2}{8Kd} \qquad 3.60$$

Equating the expressions for h in Equations (3.60) and (3.54) gives after rearrangement.

$$d = \pi L / (8 \ln L/u)$$

3.61

This value of d is used in non-steady drainage equations if D> 0.1 L.

3.4.7.2 Resistance to radial or vertically cylindrical flow as well as horizontal flow in homogeneous soil (u<D<0.3L)

Figure 3.15 shows flow to a ditch or pipe drain. If the drains do not reach the impervious layer, the flow lines will not be parallel and horizontal but will converge towards the drains (radial or vertically cylindrical flow). In this region the flow system can not be simplified to a flow field with parallel and horizontal streamlines without introducing large errors.

The radial flow causes lengthening of the flow lines. This lengthening causes a more than proportional loss of hydraulic head loss since the flow velocity in the vicinity of the drains is larger than elsewhere in the flow region. Consequently, the elevation of the water table will be higher when pipe drains replace the vertically walled ditches, the drain level remaining the same.

Schematically flow region below the drain level is subdivided into vertical, horizontal and radial flow. According to **Hooghoudt (1940)** the horizontal flow below drainage level takes place over a distance from 0.7D away from the drain to drain itself. Horizontal flow above drainage level can be ignored.

Considering Hooghoudt equation (3.21),

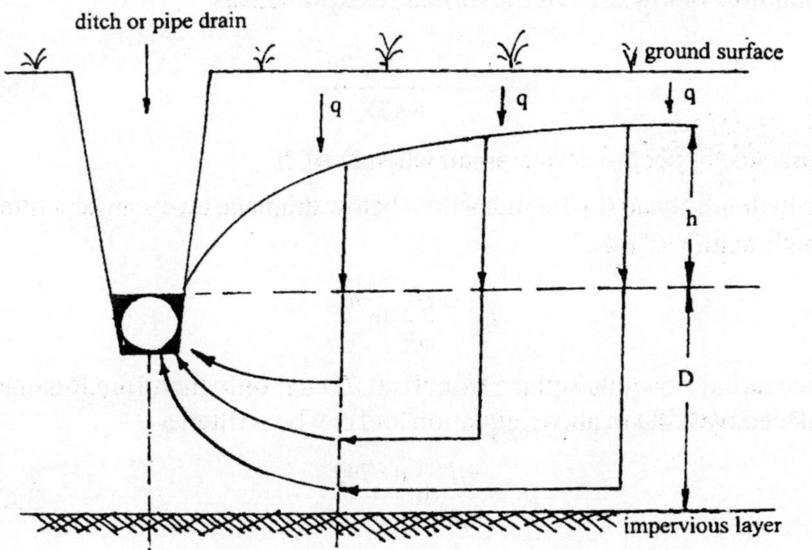

Figure 3.15. Flow to ditch or pipe drain (D<0.3L).

$$L^2 = \frac{8KDh + 4Kh^2}{q}$$

Since horizontal flow above drainage level is ignored therefore above equation takes the following form :

$$L^2 = \frac{8KDh}{q}$$

$$h = \frac{qL^2}{8KD}$$

According to Hooghoudt (1940) horizontal flow takes place only from ½ L to D√2=0.7D away from the drain, therefore in above equation L can be replaced by (L-D√2). Therefore hydraulic head (h_h) required for horizontal flow below drainage level can be expressed as:

$$h_h = \frac{q(L - D\sqrt{2})^2}{8KD} \qquad 3.62$$

In practice h_h becomes very small when D>0.1L.

The hydraulic head (h_r) for radial flow below drainage level can be written from Equation (3.14).

$$h_h = \frac{qL}{\pi K} \ln \frac{R}{r}$$

Since radial flow takes place from R=0.7D to r only therefore R can be replaced by 0.7D in above equation and can be written as.

$$h_r = \frac{qL}{\pi K} \ln \frac{0.7D}{r} \qquad 3.63$$

Hooghoudt using the method of images and superposition derived this equation. It is seen that at r=0.7D, the head $h_r = 0$. In practice h_r becomes negligibly small when $D < u = \pi r$. The total hydraulic head required for the flow below drainage level is.

$$h = h_v + h_h + h_r \qquad 3.64$$

when the hydraulic head (h_v) required to overcome the vertical resistance is negligible, Hooghoudt equation becomes,

$$h = \frac{q(L - D\sqrt{2})^2}{8KD} + \frac{qL}{\pi K}\ln\frac{0.7D}{r}$$

3.65

Above equation can also be written as,

$$h = \frac{qL}{K}F_H$$

3.66

where,

$$F_H = \frac{(L - D\sqrt{2})^2}{8DL} + \frac{1}{\pi}\ln\frac{0.7D}{r}$$

3.67

Now Hooghoudt equation reads as (from Equation 3.60)

$$h = \frac{qL^2}{8KD}$$

Equating this equation for h in Equation (3.66)

$$\frac{qL^2}{8Kd} = \frac{qL}{K}F_H$$

$$\frac{L}{8d} = F_H$$

$$d = \frac{L}{8F_H}$$

3.68

Where d is equivalent depth of impervious layer below drainage level and is a function of L, D and r. The values of d for r=0.1 m and various values of L and D are presented in **Table 3.1.**

For a homogeneous soil, with D<1/4 L and without head loss due to vertical flow, the horizontal flow above drain level can be written as,

$$h = h_h + h_r$$

$$h = \frac{qL^2}{8KD} + \frac{qL}{\pi K} \ln \frac{D}{u}$$

3.69

Corresponding to Hooghoudt equation when flow above drain level is neglected can be equated for h as below.

$$\frac{qL^2}{8Kd} = \frac{qL^2}{8KD} + \frac{qL}{\pi K} \ln \frac{D}{u}$$

by rearranging the terms

$$d = \frac{D}{1 + \frac{8D}{\pi L} \ln \frac{D}{u}}$$

3.70

This equation for equivalent depth d is presented in **Figure 3.16**, which can be referred for determination of equivalent depth of impervious layer below drain level for any value of r whereas Hooghoudt introduces an equivalent layer d to take into account the radial flow in the vicinity of drains. Van Beers (1976) takes this into account by reducing the drain spacing by a factor C, according to.

L = Lo –C

where

Lo=$[8KDh/q]^{1/2}$ is the drain spacing without radial flow, and

C = D ln (D/u), is correction term for radial flow.

Figure 3.17 can be used to determine subtraction factor C.

In order to take radial flow into account the value of d can be introduced into all equations of Hooghoudt (3.21) and (3.45). When introduced the value of d in these equations they yield:

$$q = \frac{8Kdh + 4Kh^2}{L^2}$$

3.71

Table 3.1 Equivalent depth of Hooghoudt (r=0.10 m, D and L in m).

L	5 m	7.5	10	i5	20	25	30	35	40	45	50
D											
0.5m	0.47	0.48	0.49	0.49	0.49	0.50	0.50				
0.75	0.60	0.65	0.69	0.71	0.73	0.74	0.75	0.75	0.75	0.76	0.76
1.00	0.67	0.75	0.80	0.86	0.89	0.91	0.93	0.94	0.96	0.96	0.96
1.25	0.70	0.82	0.89	1.00	1.05	1.09	1.12	1.13	1.14	1.14	1.15
1.50		0.88	0.97	1.11	1.19	1.25	1.28	1.31	1.34	1.35	1.36
1.75		0.91	1.02	1.20	1.30	1.39	1.45	1.49	1.52	1.55	1.57
2.00			1.08	1.28	1.41	1.5	1.57	1.62	1.66	1.70	1.72
2.25			1.13	1.34	1.50	1.69	1.69	1.76	1.81	1.84	1.86
2.50				1.38	1.57	1.69	1.79	1.87	1.94	1.99	2.02
2.75				1.42	1.63	1.76	1.88	1.98	2.05	2.12	2.18
3.00				1.45	1.67	1.76	1.88	1.98	2.05	2.12	2.18
3.25				1.48	1.71	1.88	2.04	2.16	2.6	2.35	2.42
3.50				1.50	1.75	1.93	2.11	2.24	2.35	2.45	2.54
3.75				1.52	1.78	1.97	2.17	2.31	2.44	2.54	2.64
4.00					1.81	2.02	2.22	2.37	2.51	2.62	2.71
4.50					1.85	2.08	2.31	2.50	2.63	2.76	2.87
5.00					1.88	2.15	2.38	2.58	2.75	2.89	3.02
5.50						2.20	2.43	2.65	2.84	3.00	3.15
6.00							2.48	2.70	2.92	3.09	3.26
7.00							2.54	2.81	3.03	3.24	3.43
8.00							2.57	2.85	3.13	3.35	3.56
9.00								2.89	3.18	3.43	3.66
10.0									3.23	3.48	3.74
0											
α	0.71	0.93	1.14	1.53	1.89	2.24	2.58	2.91	3.24	3.56	3.88

L	50	75	80	85	9	100	150	200	250
D									
0.5	0.50								
1	0.96	0.97	0.97	0.97	0.98	0.98	0.99	0.99	0.99
2	1.72	1.80	1.82	1.82	1.83	1.85	1.00	1.92	1.94
3	2.29	2.49	2.52	2.54	2.56	2.60	2.72	2.70	2.83
4	2.71	3.04	3.08	3.12	3.16	3.24	3.46	3.58	3.66
5	3.02	3.49	3.55	3.61	3.67	3.78	4.12	4.31	4.43
6	3.23	3.85	3.93	4.00	4.08	4.23	4.70	4.97	5.15
7	3.41	4.14	4.23	4.33	4.4	4.62	5.22	5.57	5.81
8	3.56	4.387	4.49	4.61	4.72	4.95	5.68	6.13	6.43
9	3.66	4.57	4.70	4.82	4.95	5.23	6.09	6.63	7.00
10	3.74	4.74	4.89	5.04	5.18	5.47	6.45	7.09	7.53
12.5		0.02	5.20	5.38	5.56	5.92	7.20	8.06	8.68
15		5.20	5.40	5.60	5.80	6.25	7.77	8.84	9 64
15.5		5.30	5.53	5.76	5.99	6.44	8.20	9.47	10.4
20			5.62	5.87	6.12	6.60	8.54	9.92	11.1
25			5.74	5.96	6.20	6.79	8.99	10.7	12.1
30							9.27	11.3	12.9
35							9.44	11.6	13.4
40								11.8	13.8
45								12.0	13.8
50								12.1	14.3
60									14.6
α	3.88	5.38	5.76	6.00	6.26	6.82	9.55	12.2	14.7

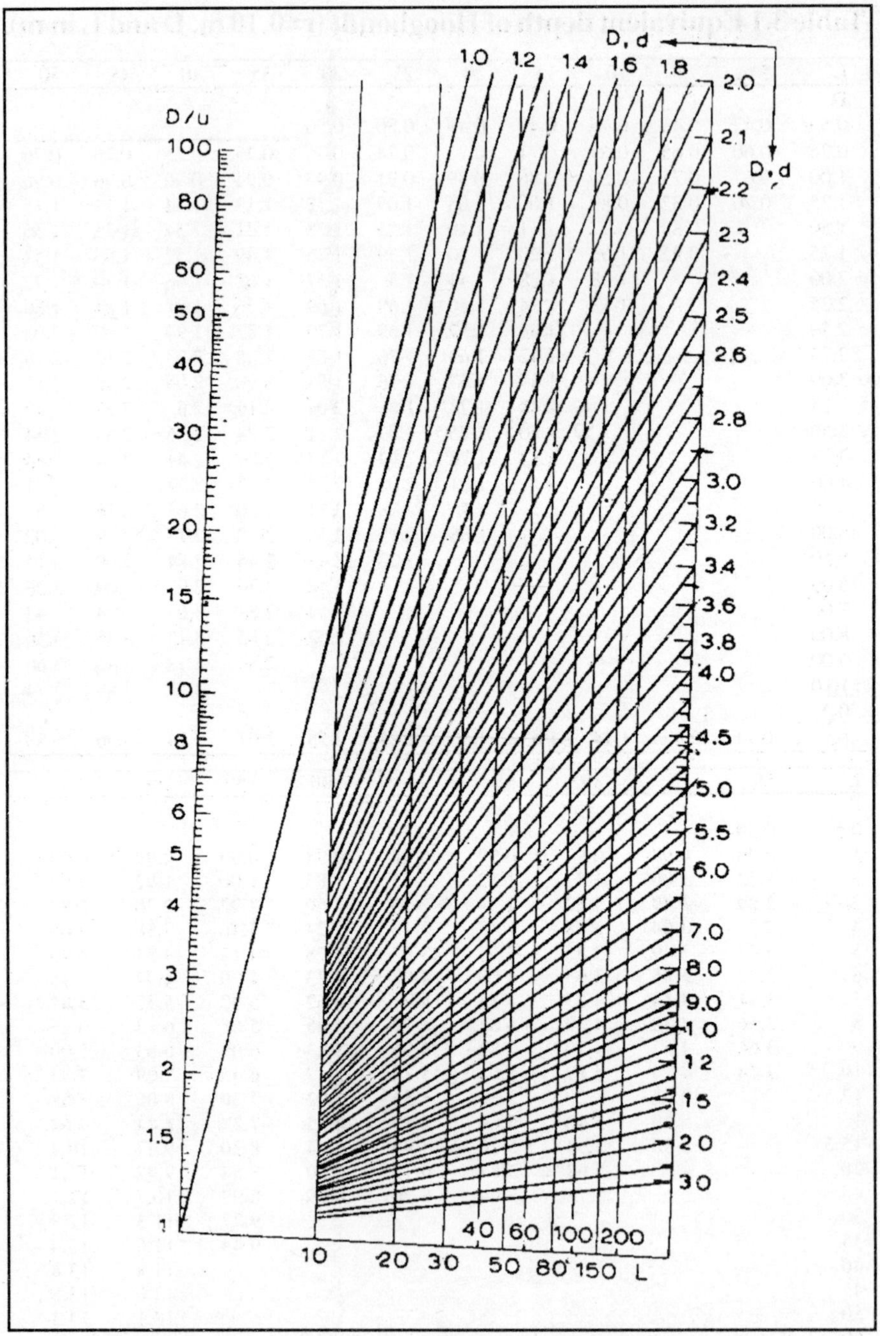

Figure 3.16. Nomograph for determination of equivalent depth.

$$q = \frac{8K_b dh + 4K_a h^2}{L^2}$$

3.72

3.4.8 Adoption of Hooghoudt's Equations for Two Layered Soils with Interface at Drainage Level

3.4.8.1 Conditions I: D>0.1L, K_a>K_b

Equation (3.45) can be used for this situation after replacing the value of D from equation by the value of d of Equation (3.61)

$$L^2 = \frac{8K_b Dh}{q} + \frac{4K_a h^2}{q}$$

(A)

$$d = \frac{\pi L}{8\ln(\frac{L}{u})}$$

(B)

Substituting this value of d in Equation A, we get.

$$L^2 = \frac{8K_b \pi L h}{q.8\ln(\frac{L}{u})} + \frac{4K_a h^2}{q}$$

$$L = \frac{\pi K_b h}{q\ln\frac{L}{u}} + \frac{4K_a h^2}{qL}$$

3.73

$$L_2 = \frac{\alpha}{\ln\frac{L_1}{u}} + \frac{\beta}{L_1} \quad 3.74$$

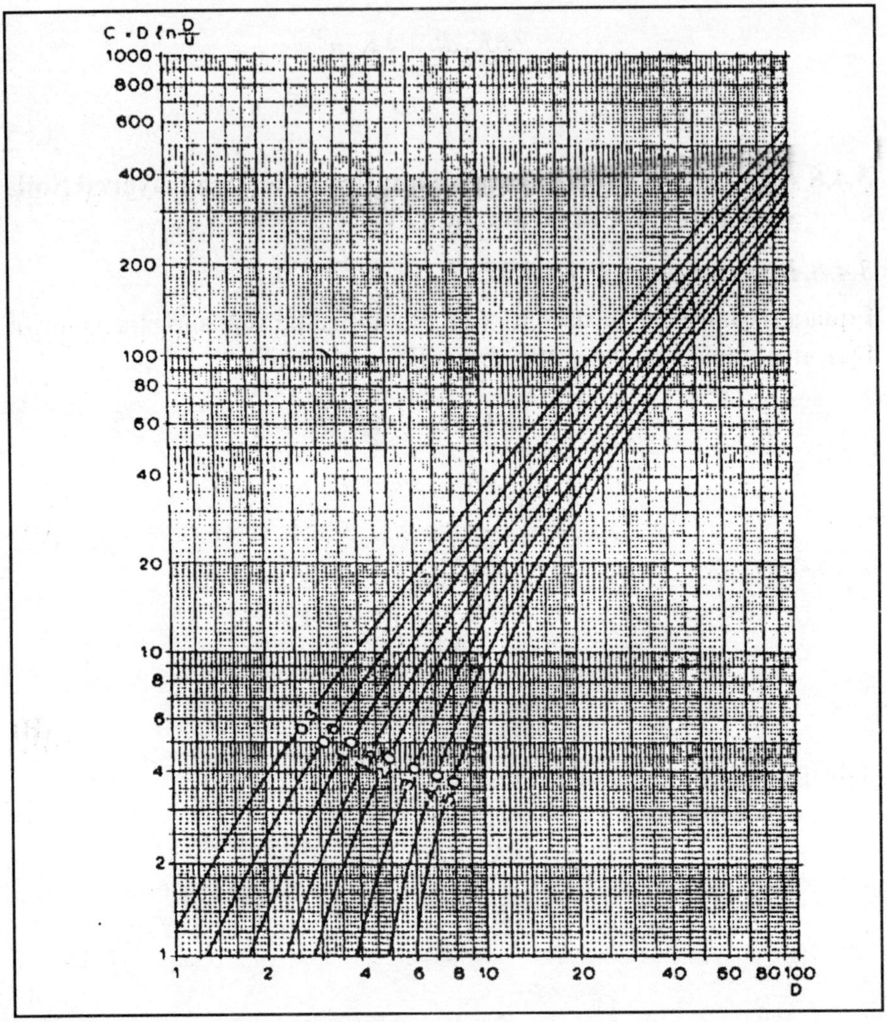

Figure 3.17. Nomograph for determination of factor C (van Beers)

where,

$L_1 = L_2 = L$

$\alpha = \pi K_b h/q$

$\beta = 4 K_a h^2/q$

Calculation can be made following the following steps:

1. Assume L_1;

2. Calculate L_2 from Equation (3.74)
3. If $L_2>L_1$, increase L_1 and repeat the procedure until $L_2 \cong L_1$;
4. If $L_2<L_1$, reduce the L_1, and repeat the procedure unless $L_2=L_1$

3.4.8.2 Conditions II: D>0.1L, $K_a<K_b$

For such condition hydraulic head available for horizontal flow can be taken from Equation (3.49)

$$h_h = h\,(1\text{-}q/K_a)$$

Hooghoudt's equation when resistance to vertically cylindrical flow is dominating ($K_a<K_b$) reads,

$$h = qL/\pi K_b \ln L/u$$

but only h_h is available for horizontal flow therefore h can be replaced by h_h

$$h(1\text{-}q/K_a) = qL/\pi K_b \ln L/u$$

Which can also be written as,

$$L \ln L/u = \pi K_b h(1/q - 1/K_a)$$

$$\text{Or } L \ln L/u = y$$

where,

$$y = \pi K_b h(\frac{1}{q} - \frac{1}{K_a}) \tag{3.75}$$

Above equation can be given following form

$$L = \alpha y^{\beta} \tag{3.76}$$

where,

$\alpha = 0.56\ u^{0.29}$

$\beta = 0.85 - 0.046 \log u$

3.4.8.3 Conditions III: D<0.3L, $K_a > K_b$

Hooghoudt equation for two layered soil when drainage level is at interface reads,

$$L^2 = \frac{8K_b D.h + 4K_a h^2}{q}$$

Hydraulic head for horizontal flow (h_h) can be determined using the above equation when horizontal flow above the drainage level and below the drainage level both are considered. Expression for h_h can be written as,

$$h_h = \frac{qL^2}{8K_b D + 4K_a h}$$

Since horizontal flow takes place only up to a distance of (L-D√2) from the midway of the drain therefore in the above equation L can be replaced by (L-D√2), then above equation becomes as follows.

$$h_h = \frac{q(L - D\sqrt{2})^2}{8K_b D + 4K_a h}$$

Hydraulic head for radial flow (h_r) can be calculated by using Equation (3.63), which is being reproduced below.

$$h_r = \frac{qL}{\pi K_b} \ln \frac{0.7D}{r}$$

For total head, hydraulic head for horizontal flow (h_r) and hydraulic head for radial flow (h_r) need to be added as follow.

$$\frac{8K_b Dh + 4K_a h^2}{q} = L^2 + 2D^2 - 2DL\sqrt{2} + \frac{L(8K_b D + 4K_a h)}{\pi K_b} \ln \frac{0.7D}{r}$$

$$\frac{8K_b Dh + 4K_a h^2}{q} = L^2 + 2D^2 - 2DL\sqrt{2} + \frac{L(8K_b D + 4K_a h)}{\pi K_b} \ln \frac{0.7D}{r}$$

$$\frac{4K_bDh + 2K_ah^2}{q} - D^2 = \frac{1}{2}L^2 + (\frac{4K_bD + 2K_ah}{\pi K_b}\ln\frac{0.7D}{r} - D\sqrt{2})L$$

or

$$\beta = ½\,L^2 + \alpha\,L$$

3.77

where,

$$\alpha = \frac{4K_bD + 2K_ah}{\pi K_b}\ln\frac{0.7D}{r} - D\sqrt{2}$$

3.78

$$\beta = \frac{4K_bDh + 2K_ah^2}{q} - D^2$$

3.79

If soil is homogeneous earlier equation may be also written as,

$$\frac{4KDh + 2Kh^2}{q} - D^2 = \frac{1}{2}L^2 + L[\frac{(4KD + 2Kh)}{\pi K}\ln\frac{0.7D}{r} - D\sqrt{2}]$$

or

$$\frac{2Kh(2D + h)}{q} - D^2 = \frac{1}{2}L^2 + L[\frac{(4D + 2h)}{\pi}\ln\frac{0.7D}{r} - D\sqrt{2}]$$

or

$$\beta = ½\,L^2 + \alpha\,L$$

where,

$$\beta = \frac{2Kh(2D + h)}{q} - D^2$$

$$\alpha = (\frac{4D+2h}{\pi})\ln\frac{0.7D}{r} - D\sqrt{2}$$

3.4.8.4 Conditions IV: $D<0.3L$, $K_a<K_b$

When resistance to vertical flow is dominating the head available for horizontal flow is equal to,

$$h_h = h - h_v = h\,(1 - q/K_a)$$

Therefore this value of h_h is placed in place of h as follow.

$$\frac{(8K_bD+4K_ah)h}{q} - 2D^2 = L^2 - 2DL\sqrt{2} + \frac{L(8K_bD+4K_ah)}{\pi K_b}\ln\frac{0.7D}{r}$$

or

$$\frac{(8K_bD+4K_ah)h}{q}(1-\frac{q}{K_a}) - 2D^2 = L^2 + 2L(\frac{4K_bD+2K_ah)}{\pi K_b}\ln\frac{0.7D}{r} - D\sqrt{2})$$

or

$$h(4K_bD+2K_ah)(\frac{1}{q}-\frac{1}{K_a}) - D^2 = \frac{1}{2}L^2 + L(\frac{L(4K_bD+2K_ah)}{\pi K_b}\ln\frac{0.7D}{r} - D\sqrt{2}$$

$$\beta = \frac{1}{2}L^2 + \alpha L$$

where,

$$\alpha = \frac{K_bD+2K_ah}{\pi K_b}\ln\frac{0.7D}{r} - D\sqrt{2} \qquad 3.80$$

$$\beta = (4K_bD+2K_ah)h(\frac{1}{q}-\frac{1}{K_a}) - D^2 \qquad 3.81$$

3.4.8.5 Adoption of Spacing Equation for Multi-layered or Anisotropic Soils

I. Homogeneous anisotropic soils

When horizontal hydraulic conductivity (K_h) and vertical hydraulic conductivity (K_v) are different in a soil but constant anywhere in the soil then soil is called as homogeneous anisotropic soil.

Multi-layered soils can be treated as homogeneous anisotropic soil by approximation. Drain spacing equations with drains on an impermeable layer can be adopted as follows.

1. Calculate the maximum depth of flow (H_m) with the help of Equation (3.39). First assume a certain value of L and later on assumed value of L is checked by calculated value of L by spacing equation. If difference is large a new value of L is assumed and checked by calculated value. By such trial and error method drain spacing is calculated by minimizing the difference in assumed and calculated value of spacing.
2. If $H_m < H_o$, replace the H_o value by H_m ,but if $H_m > H_o$ the H_o value ought to be maintained as it is. Say, the applicable value H'. ($H' = H_o$ if $H_m > H_o$ and $H' = H_m$ if $H_m < H_o$).
3. Calculate the hydraulic head h_v to overcome resistance due to vertical flow by following equation.

$$h_v = q\frac{(H'+h)}{K_v}$$

4. Subtract the head loss h_v from hydraulic head to get

$$h' = h - h_v = h - \frac{q(H'+h)}{K_v} \qquad 3.82$$

where,

h' = hydraulic head available for horizontal flow.

Now Equation (3.50) to (3.52) can be applied using $h_h = h'$ instead of h, and K_h instead of K.

Spacing equation with radial components can be adopted as follows:

The hydraulic conductivity to be used in the radial flow component has to be calculated as the geometric mean of K_v and K_h as,

$$K_g = (K_h.K_v)^{1/2} \tag{3.83}$$

Now the maximum depth of flow (H_m) can be calculated according to following procedure. Refer **Figure 3.9** when drain ditch is resting on impervious layer. The drain spacing equation for such situation can be written considering horizontal and radial flow as well as vertical flow.

Head loss due to vertical flow, $h_v = q\,(Ho+h)/K_v$ from Equation (3.34).

The head loss due to horizontal flow, $h_v = q(H_o+h)/K_v$ from Equation (3.62) and head loss due to radial flow, $h_r = (q/\pi K_r \ln H_o/u)$ from Equation (3.63) where, $K_r = K_g = (K_h K_v)^{1/2}$

$$h = h_v + h_h + h_r = \frac{q(H_o + h)}{K_v} + \frac{q(L - D\sqrt{2})^2}{8K_h D} + \frac{qL}{\pi K_g}\ln\frac{0.7D}{r}$$

$D\sqrt{2}$ is negligible as compared to L, therefore,

$$h = \frac{q(H_o + h)}{K_v} + \frac{qL^2}{8K_h D} + \frac{qL}{\pi K_g}\ln\frac{0.7D}{r} \tag{3.84}$$

when ditch is at impervious layer D can be replaced by H_o. Now above equation reduces to,

$$h = \frac{q(H_o + h)}{K_v} + \frac{qL^2}{8K_h H_o} + \frac{q}{\pi K_g}(\ln H_o + \ln\frac{0.7D}{r}) \tag{3.85}$$

$$\frac{h}{q} = \frac{(H_o + h)}{K_v} + \frac{L^2}{8K_h H_o} + \frac{L}{\pi K_g}.(\ln H_o + \ln \frac{0.7D}{r}) \quad 3.86$$

Differentiating the above equation as:

$$\frac{1}{q}\frac{dh}{dH_o} = 0$$

for $H_o = H_m$

$$\frac{1}{q}\frac{dh}{dH_o} = \frac{1}{K_v} + (-\frac{L^2}{8KH_o}) + \frac{L}{\pi K_g}.\frac{1}{H_o}$$

$$0 = \frac{1}{K_v} - \frac{L^2}{8KH_o} + \frac{L}{\pi K_g}.\frac{1}{H_m}$$

with some mathematical play above equation can also be written as,

$$8\pi K_h K_g H_m^{\ 2} + 8K_v K_h L H_m - \pi K_v K_g L^2 = 0$$

or

$$\alpha H_m^{\ 2} + \beta H_m - \gamma = 0$$

Therefore,

$$H_m = \frac{-\beta + \sqrt{\beta^2 + 4\alpha.\gamma}}{2\alpha} \quad 3.87$$

where,

$\alpha = 8\pi K_h K_g$

$\beta = 8 K_v K_h L$

$\gamma = \pi K_v K_g L^2$

3.4.9 Application of Hooghoudt Drain Spacing Equation

Hooghoudt equation can be used for drain spacing calculation, if the factor q, h, K, D and r are known. The same formula can also be used for calculation of soil parameters such as, K and D, provided q, h, L and r are known. Since drain spacing L depends on the equivalent depth of impervious layer below drainage level, d, which in turn is a function of drain spacing and cannot be expressed explicitly in drain spacing L. Therefore, its use for drain spacing calculation is based on trial and error. To avoid trial and error method nomographs may be used for drain spacing calculation as shown in **Figure 3.18** and **3.19**.

Example 3.1

For the drainage of a canal irrigated area, a drain pipe having 0.10 m radius will have to be used. The drains will be placed at a depth of 1.80 m below the ground surface. A relatively semi-impervious layer was found at a depth of 6.8 m below the ground surface. Average saturated hydraulic conductivity of the area was estimated to be 0.80 m/day using auger hole technique. Irrigation is applied once in 20 days in the area. The average irrigation losses amount to be 0.002 mm/day. What drain spacing must be applied when average water table depth of 1.20 m below soil surface is to be maintained? Fluctuations in water table may be assumed small.

Solution 3.1

Given:

$D = 6.8 \text{ m} - 1.8 \text{ m} = 5.0 \text{ m}$

$h = 0.60 \text{ m}$

$K = 0.80 \text{ m/day}$

$q = 0.002 \text{ m/day}$

$r = 0.10 \text{ m}$

Information available in the question may be transformed to a diagram as shown below.

Substituting above data in Equation (3.21) one will obtain,

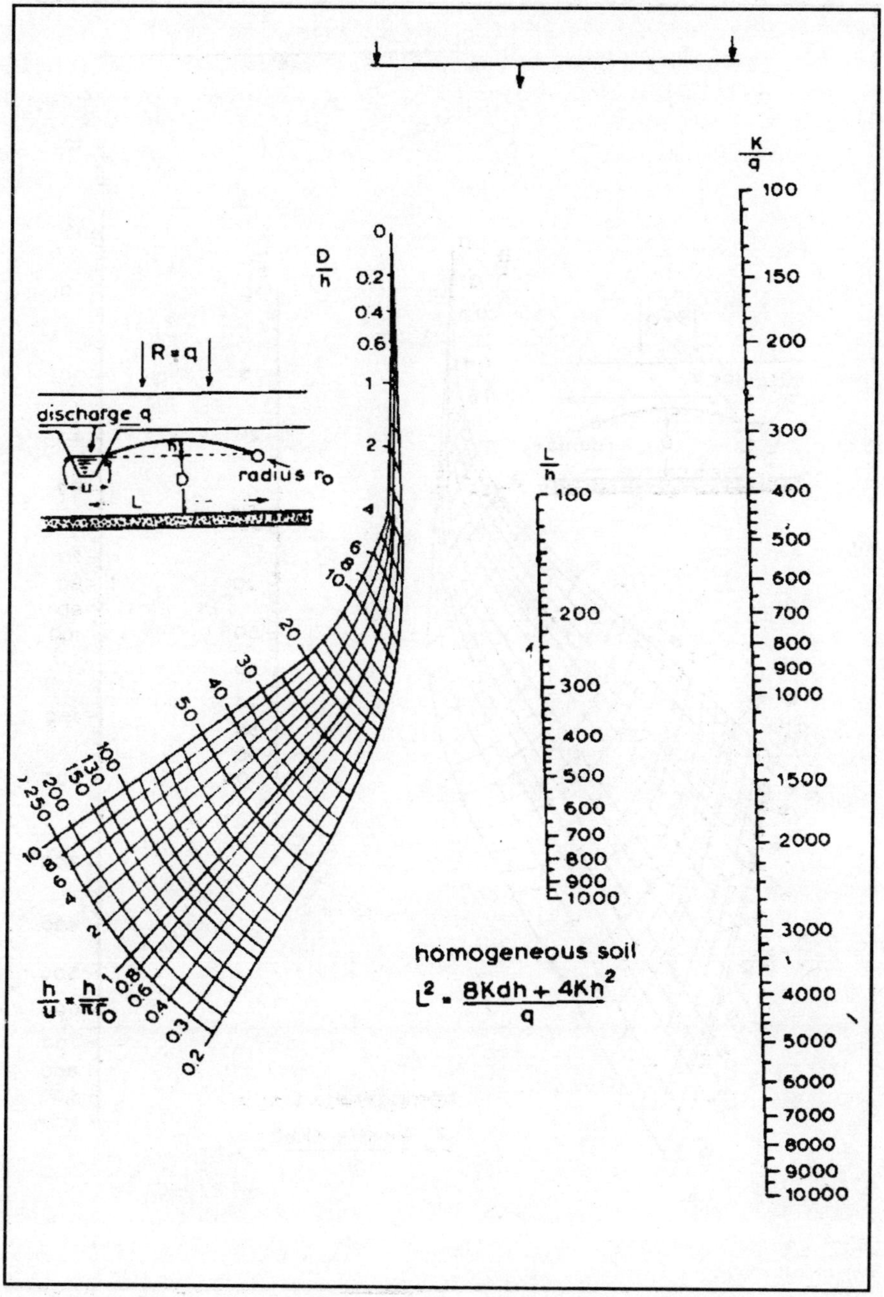

Figure 3.18. Nomograph for determination of drain spacing (L/h<100) (**Boumans**,1963)

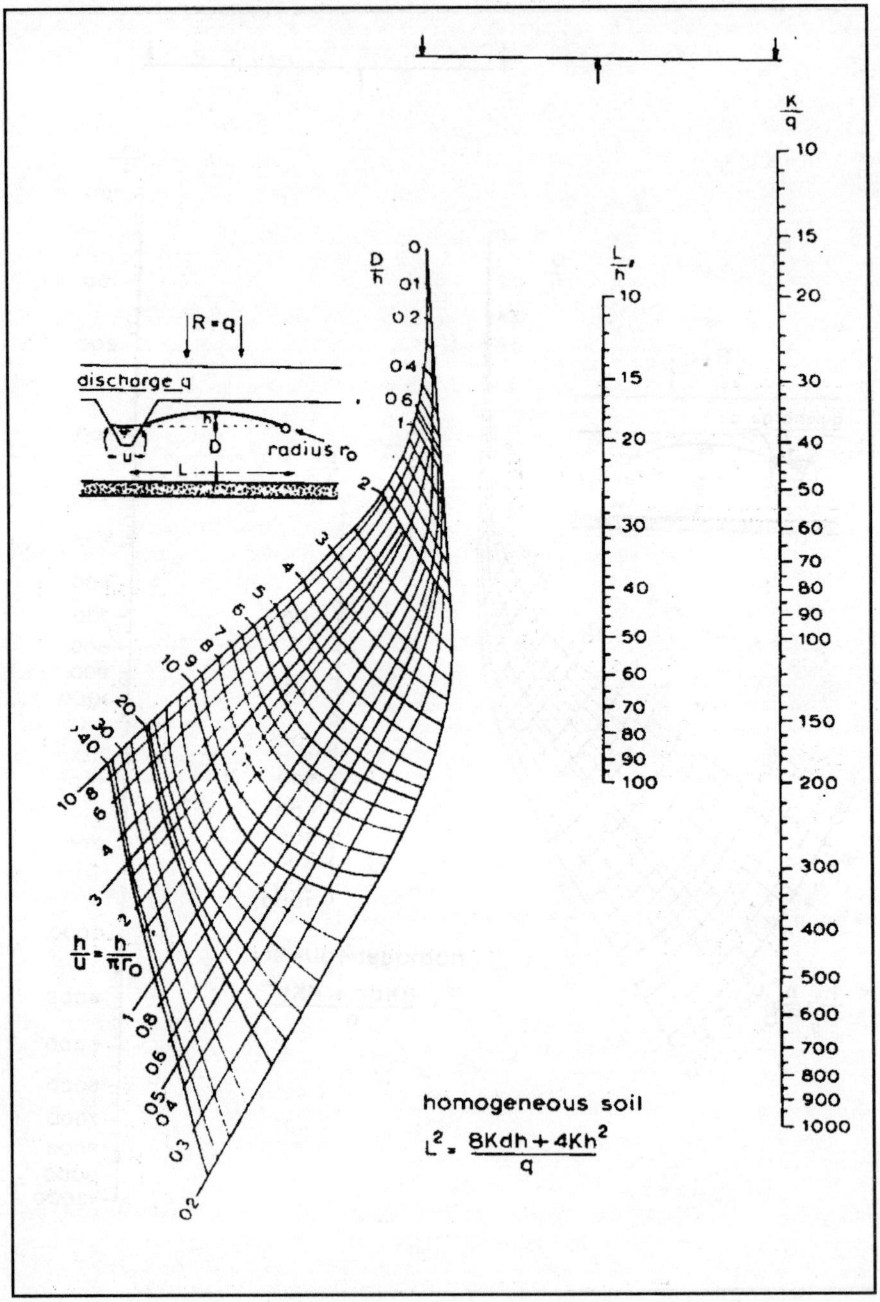

Figure 3.19. Nomograph for determination of drain spacing (L/h>100) (**Boumans**,1963)

$$L^2 = \frac{8KDh + 4Kh^2}{q} = \frac{8X0.8X5X0.6 + 4X0.8X0.6^2}{0.002}$$

$L^2 = 10176$

L = 100.9 m

To account radial flow Equation (3.71) can be used as follows,

$$L^2 = \frac{8Kdh + 4Kh^2}{q} = \frac{8X0.8XdX0.6 + 4X0.8X0.6^2}{0.002}$$

$L^2 = 576 + 1920\ d$

Trial 1:

Assume L = 80 m and read from **Table 3.1**; d = 3.55 m

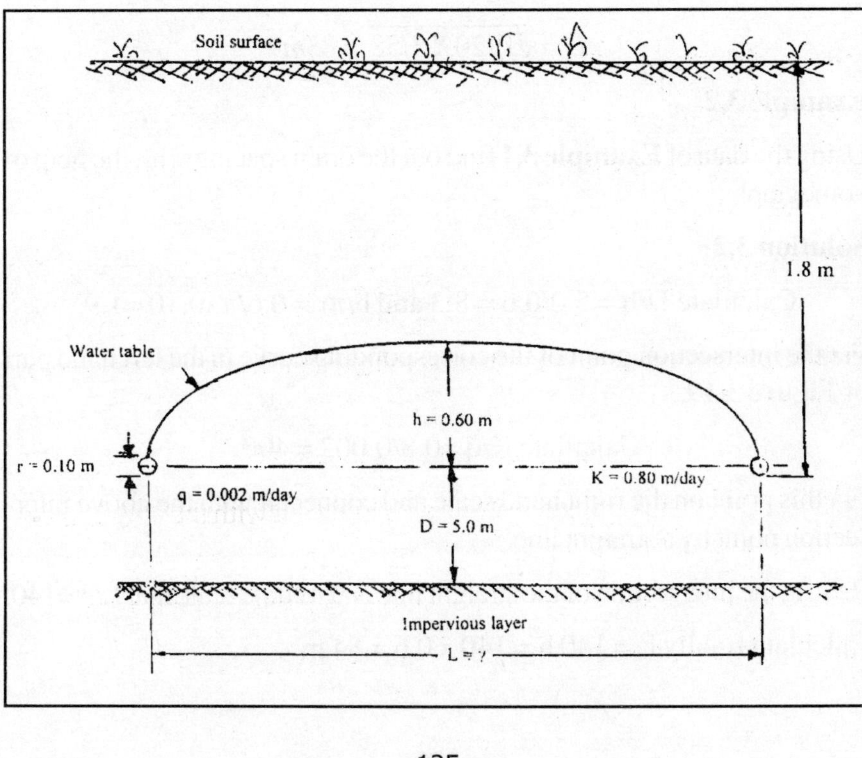

$L^2 = 1920\ d + 576 = 1920 \times 3.55 + 576 = 7392\ m^2$

This is not in agreement with $L^2 = 6400$ (assumed value).

Therefore L = 80 m is apparently too small.

Trial 2:

Assume L = 87 m and read value of d from **Table 3.1** as d = 3.63 m

$L^2 = 1920\ d + 576 = 1920 \times 3.63 + 576 = 7546\ m^2$

This is sufficiently close to assumed value of $L^2 = 7569$

Therefore conclusion can be made that 87 m drain spacing satisfy the above condition and can be adopted.

If the flow above drain level is neglected as the term 576 which represent the flow above drain level is comparatively small. Therefore,

$$L = \sqrt{1920d}$$

After making some trials and adopting d = 3.58 m

$$L = \sqrt{1920 X 3.58} = 83m$$

Example 3.2

Using the data of **Example 3.1** find out the drain spacing with the help of nomographs.

Solution 3.2

Calculate $D/h = 5.0/0.6 = 8.3$ and $h/\pi r = 0.6/\pi x 0.10 = 1.9$.

Fix the intersection point of the corresponding curve in the left hand part of **Figure 3.19**.

Calculate $K/q = 0.8/0.002 = 400$.

Fix this point on the right hand scale and connect it with the above intersection point by a straight line.

Read at the intersection of the straight line and middle scale that $L/h = 140$.

Calculate finally $L = 140\ h = 140 \times 0.6 = 84$ m.

Example 3.3

Calculate the drain spacing using the following set of data for different conditions.

Average drain discharge q = 0.003 m/day

Average depth of water table midway between the drain = 1.0 m

The drainage level can be taken at a1.5 m depths.

Drain radius r = 0.05 m but it is to be surrounded by a gravel envelope of 5 cm thickness.

Average radius for ditch = 0.63 m.

Condition I : When soil is uniform of 1.5 m depth, K = 2 m/day

Condition II : When soil is uniform of 10 m depth, K = 2 m/day

Condition III: When soil is uniform of 100 m depth, K = 2 m/day

Solution 3.3

Condition I:

Since depth of soil is 1.5 m and drainage level is also 1.5 m which shows drainage level is at impervious layer, for such condition Equation (3.44) can be used.

$$L = 2h\sqrt{(K/q)}$$

h = depth of drainage level – depth of water table below ground surface at midpoint of the drain.

$$h = 1.5 - 1.0 = 0.5 \text{ m}$$

Therefore, $L = 2 \times 0.5 \sqrt{(2/0.003)}$

$= 26$ m.

Condition II:

For such condition Equation (3.77) can be used which is,

$$\beta = \frac{1}{2}L^2 + \alpha L$$

For homogeneous soil,

$$\beta = \frac{2Kh(2D+h)}{q} + D^2$$

$$\alpha = (\frac{4D+2h}{\pi})\ln\frac{0.7D}{r} - D\sqrt{2}$$

Depth of impermeable layer below drainage level D = Depth of soil – Depth of drainage level

D = 10 m – 1.5 m = 8.5 m

h = 1.5 m – 1.0 m = 0.5 m

K = 2 m/day

Average drain radius is to be used because drain is surrounded by a 5 cm thick gravel filter.

$\therefore$ r = 0.05 + 0.05 = 0.10 m

$$\beta = \frac{2X2X0.5(2X8.5+0.5)}{0.003} - 8.5^2$$

$$\beta = 11593$$

$$\alpha = \frac{4X8.5+2X0.5}{3.14}\ln\frac{0.7X8.5}{0.10} - 8.5\sqrt{2}$$

$$\alpha = 34$$

$$\therefore L = -\alpha + \sqrt{\alpha^2 + 2\beta}$$

$$L = -34 + \sqrt{34^2 + 2X11593}$$

$$= 122m$$

Check D< 0.3 L = 8.5 < 0.3 x 122

= 8.5 < 36.6

Ditch with average radius r = 0.63 m

$$\alpha = \frac{4D + 2h}{\pi} \ln \frac{0.7D}{u}$$

$$\alpha = \frac{4X8.5 + 2X0.5}{\pi} \ln \frac{0.7X8.5}{0.63}$$

$$\alpha = 13$$

$$\beta = 11593$$

$$\therefore L - 13 + \sqrt{13^2 + 2X11593} = 140m$$

Condition III:

When depth of impermeable layer is at greater depth (D>0.1 L) the radial flow is dominating for such condition Equation (3.57) can be used which is reproduced below.

$$L = \alpha y\beta$$

with $\alpha = 0.56\, u^{0.29}$

$\beta = 0.85 - 0.46 \log u$

$y = \pi\, Kh/q$

For ditch drain

$r = 0.63$ m

$K = 2$ m/day

$\alpha = 0.56\, u^{0.29}$

$= 0.56\, (\pi \times 0.63)^{0.29} = 0.68$

$\beta = 0.85 - 0.46 \log (\pi \times 0.63)$

$= 0.84$ m

$y = \pi \times 2\, (1.5 - 1)\, /0.003 = 1047$

Hence,

$L = \alpha\, y^{\beta} = 0.68 \times 1047^{0.84}$

$= 234$ m

Check: D > 0.1L

(100-1.5) > 0.1 x 234

98.5>23.4

3.4.10 Ernst Drain Spacing Equation

The Ernst steady state drain spacing formula is applicable to two layered soils. It offers an improvement on the former formulas between the two layers, lower layer can be either above or below drain level. It is especially useful when the upper layer has a considerably lower hydraulic conductivity than the lower layer.

Like the Hooghoudt equation, the Ernst equation is found as the sum of the hydraulic heads required for various flow components in which the flow towards the drain may be schematically divided.

In analogy with Ohm's law, we may write for ground water flow.

$$q = h/w \text{ or } h = qw$$

where,

q = the flow rate, [L/T]

h = hydraulic head, [H]

w = resistance to flow, [T]

In this way if we divide the flow towards the drains into vertical, horizontal and radial flow, the total hydraulic head may be given by;

$$h = h_v + h_h + h_r = q\, W_v + q\, L\, W_h + qL\, W_r \qquad 3.88$$

where subscript v, h and r refer to vertical, horizontal and radial flow. The horizontal and radial flow equal qL i.e. the drain discharge per unit length of drain, whereas vertical flow equals to q, the drain discharge per unit surface area.

The hydraulic head required to overcome the resistance to vertical flow h_v from Equation (3.3) (replacing v by q and L by D_v).

$$h_v = q\frac{D_v}{K_v}$$

3.89

The horizontal hydraulic head required to overcome the resistance due to horizontal flow from Equation (3.35) neglecting flow above drainage level can be written as below.

$$h_h = \frac{qL^2}{8\Sigma(KD)_h}$$

3.90

And hydraulic head required to overcome the resistance due to radial flow can be worked out from Equation (3.14) by replacing R with a.D_r as follows :

$$h_r = \frac{qL}{\pi K_r}\ln\frac{aD_r}{u}$$

3.91

Therefore, the total hydraulic head required for flow towards drain can be written as below.

$$h = q\frac{D_v}{K_v} + q\frac{L^2}{8\Sigma(KD)_h} + q\frac{L}{\pi K_r}\ln\frac{aD_r}{u}$$

3.92

where,

h = total hydraulic head or water table height midway between the drains, [L]

q = drain discharge per unit surface area, [L/T]

L = drain spacing, [L]

K_r= saturated hydraulic conductivity in the layer with radial flow, [L/T]

K_V= saturated hydraulic conductivity for vertical flow, [L/T]

D_v = thickness of layer over which vertical flow is considered, [L]

D_r= thickness of layer in which radial flow is considered, [L]

$\Sigma(KD)_h$ = transmissivity of the soil layers through which horizontal flow is considered, $[L^2/T]$

u = wetted perimeter of the drain, [L]

a = geometry factor for radial flow depending on the flow conditions.

Which is the **Ernst (1956, 1962)** steady state drain spacing formula. **Figures 3.20 (a-d)** describe the different flow geometry. The values of D_v, $\Sigma(KD)_h$, Dr, a and u are to be determined in accordance with the soil profile and the relative position and size of the drains. The approximate values are derived form the following data, which characterize the specific drainage conditions, namely.

D_1 = average thickness below the water table of upper layer with hydraulic conductivity K_1, [L]

D_2 = thickness of lower layer with hydraulic conductivity K_2, [L]

D_o = thickness below drain level of the layer in which drain is located, [L]

h = water table height above drain level at mid point, [L]

y = water table depth in drain, [L], for pipe drain y=0.

Vertical flow takes place in the layer between the maximum water table height midway between the drains and the drain bottom. Usually the thickness of the layer for vertical flow can be taken as $D_v = y + h$ for ditches and Dv = h for pipe drains. In fact this should be ½ (y+h) and ½ h respectively, but usually this factor is of little importance.

Horizontal flow occurs over the whole thickness of the aquifer, thus $\Sigma(KD)_h = K_1D_1 + K_2D_2$. If the depth to impervious layer increases, the value of K_2D_2 increases too, making $\Sigma(KD)_h$ tending to infinity and horizontal resistance to zero.

In order to prevent this the total thickness of layers below the drains D_o or D_o+D_2 is restricted to ¼ L when impermeable layer is deeper than ¼ L below drain level.

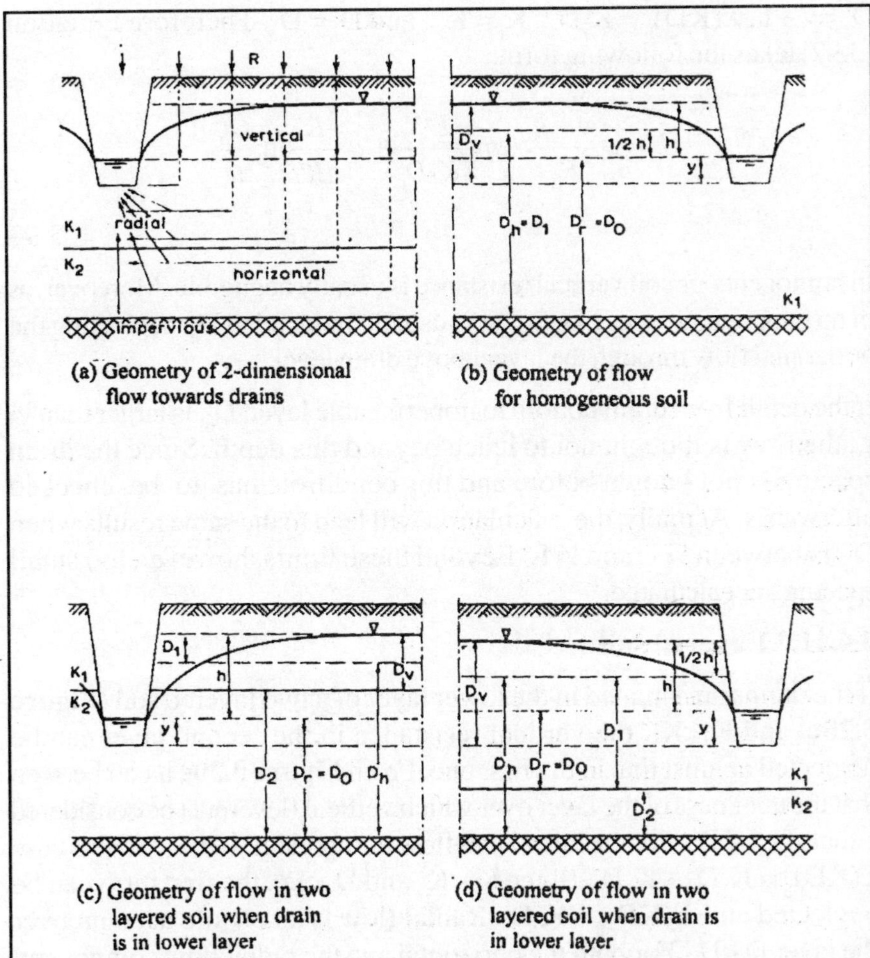

Figure 3.20. Flow geometry for Ernst equation.

Radial flow is taken into account only in the layer below drain level, thus $D_r = D_o$, with the condition that for radial flow the same restriction should be applied for D_o as for horizontal flow, $D_o < 1/4\,L$.

3.4.11 Adoption of Ernst Equation for Different Conditions

3.4.11.1 Homogeneous soil

In a homogeneous soil (**Figure 3.20b**, $D_2 = 0$) the value of a = 1,

D_2 =y + h, $\Sigma(KD)_h = K_1D_1$, $K_r = K_1$, and $D_r = D_o$. Therefore Equation (3.92) takes the following form.

$$h = q\frac{y+h}{K_1} + q\frac{L^2}{8K_1D_1} + q\frac{L}{\pi K_1}\ln\frac{D_o}{u}$$

3.93

In homogeneous soil vertical resistance is usually negligible. Moreover, as in most practical cases $h \ll D_o$, D_1 is usually reduced to D_o, neglecting the horizontal flow through the layer above drain level.

If the depth from drain bottom to impermeable layer, D_o, is larger than ¼ L, the flow is thought not to reach beyond this depth. Since the drain spacing is not known before and this condition has to be checked afterwards. Actually, the calculation will lead to the same results when D_o, is between ¼ L and ½ L. Beyond these limits, however, too small spacing are calculated.

3.4.11.2 Layered soil

If the drains are situated in the lower layer of a two layered soil (**Figure 3.20c**) and $K_1<K_2$, the vertical resistance in the second layer can be neglected against that in the first one. From **Figure 3.20c** it can be seen that the thickness of the layer over which vertical flow must be considered equals $D_v = 2D_1$. For the horizontal flow component we have in this case $\Sigma(KD)_h = K_1D_1 + K_2D_2$. Since $K_1<K_2$ and $D_1<D_2$, the first term can be neglected and $\Sigma(KD)_h = K_2D_2$. Radial flow is taken into account over the layer $D_r<D_o$. For both the horizontal and the radial flow component, again the restriction is made that the thickness D_o may not exceed ¼ L. The equation then becomes.

$$h = q\frac{2D_1}{K_1} + q\frac{L^2}{8K_2D_2} + q\frac{L}{\pi K_2}\ln\frac{D_o}{u}$$

3.94

If the drain is entirely in the upper layer of a two layered soil (**Figure 3.20d**) the following conditions must be discerned with respect to the geometry factor a.

I. $K_2 > 20\ K_1$

The geometry factor a = 4 and Equation (3.92) becomes.

$$h = q\frac{y+h}{K_1} + q\frac{L^2}{8(K_1D_1 + K_2D_2)} + q\frac{L}{\pi K_1}\ln\frac{4D_o}{u}$$

3.95

II. $0.1K_1 > K_2 < 20\ K_1$

The geometry factor a has to be determined from the nomograph given in **Figure 3.21** and to be introduced in Equation (3.92)

III. $0.1K_1 > K_2$

The geometry factor a =1. The lower layer can be considered impervious and case reduces to a homogeneous soil underlain by an impervious boundary, so that Equation (3.93) is applicable.

The wetted perimeter u for ditch drain can be calculated as:

$$u = b + 2y\sqrt{(s^2+1)}$$

3.96

where,

b = bottom width of ditch, [L]

y = water depth in the ditch, [L]

s = side slope of the ditch, horizontal/ vertical

It is very difficult to work out wetted perimeter of pipe drains laid in trenches and surrounded by enveloping material of good permeability. Under normal conditions u is determined as:

$$u = b + 2 \times 2r$$

3.97

where,

b = width of trench, [L]

r = drain radius, [L]

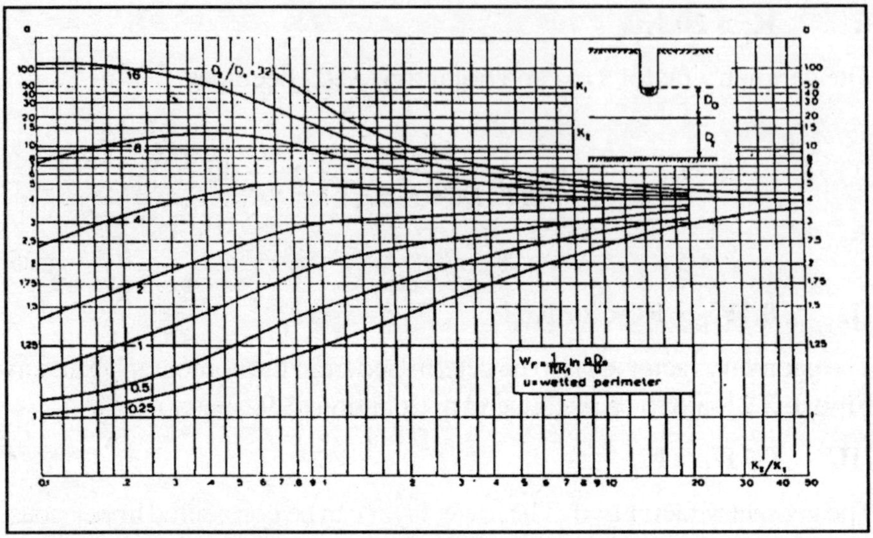

Figure 3.21. Nomograph for determination of the gemetry function "a" for radial resistance (**van Beers**, 1965)

If the filter material is used, it is advisable to replace 2r by the height of the filter.

Ernst (1978) also developed drain spacing equations for soil consisting of three different soil layers with interface at drainage level and below drainage level with $u<D1<0.3L$ (**Figure 3.23**).

In many river plains and delta areas the hydraulic conductivity increases with the depth. It is specially in these conditions Ernst equation is useful. Ernst equation for such conditions reads:

$$h = q\frac{y}{K_a} + q\frac{L^2}{8K_1D_1} + q\frac{L}{\pi K_1}\ln\frac{aD_1}{u} \qquad 3.98$$

where,

K_a = saturated hydraulic conductivity of layers above drain level, [L/T]

$K_1D_1+K_2D_2$ = transmissivity of the layers lying below the drains level, [L^2/T]

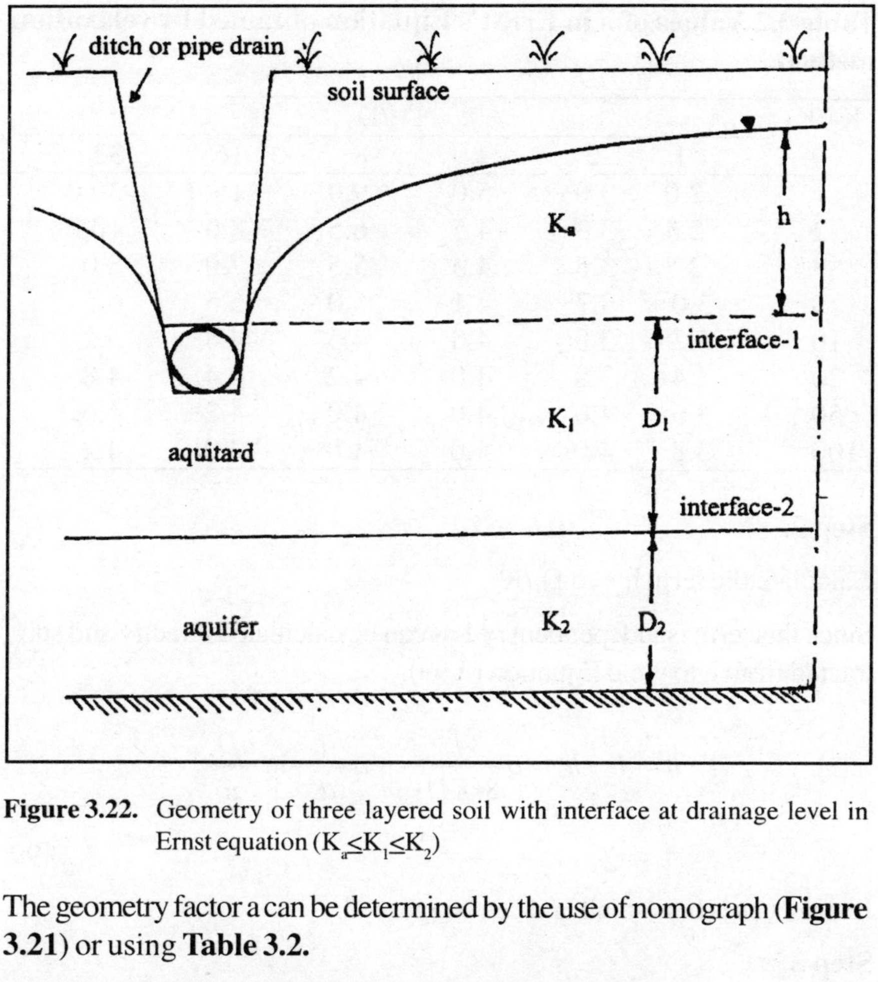

Figure 3.22. Geometry of three layered soil with interface at drainage level in Ernst equation ($K_a \leq K_1 \leq K_2$)

The geometry factor a can be determined by the use of nomograph (**Figure 3.21**) or using **Table 3.2.**

3.4.12 Application of Ernst drain spacing equation

The drain spacing can be calculated directly or with the help of the nomograph given in **Figure 3.21 and 3.23 (van Beers, 1965)**. The computation is carried out in steps to facilitate the right choice of equation.

Step 1.

Check the soil profile.

If the soil is homogeneous or if the depth of the layer in which the drain will be situated is more than ¼ L, Equation (3.93) can be used. If less, go to step 2 and 3.

Table 3.2 Values of a in Ernst's Equation obtained by relaxation method.

K_2/K_1	D_2/D_1					
	1	2	4	8	16	32
1	2.0	3.0	5.0	9.0	15.0	30.0
2	2.4	3.3	4.5	6.5	8.0	10.0
3	2.7	3.5	4.3	5.5	7.0	8.0
5	3.0	3.7	4.1	5.0	5.5	6.5
10	3.2	3.8	4.0	4.5	4.8	5.2
20	3.4	3.8	4.0	4.2	4.4	4.8
50	3.6	4.0	4.0	4.0	4.2	4.6
100	3.8	4.0	4.0	4.0	4.0	4.2

Step 2.

Calculate the term $h_v = q\,D_v/K_v$

Since this term is independent of L it can be calculated directly and subtracted from h to yield Equation (3.99).

$$h' = h - h_v = q\frac{L^2}{8(KD)_h} + q\frac{L}{\pi K_r}\ln\frac{aD_r}{u} \tag{3.99}$$

In most cases h_v is small and can be ignored.

Step 3.

Determine the geometry factor 'a'.

If K_2>20K_1, set a=4 and apply Equation (3.95).

If 0.1K_1>K_2<20K_1, determine 'a' from **Figure 3.21** and apply Equation (3.99).

If K_2>0.1K_1 , set a=1 and consider soil homogeneous and apply Equation (3.93).

If drainage level is at the interface, use Equation (3.98).

Example 3.4

For a tile drainage system, drain pipes with a radius of 0.1 m in a trench of width 0.25 m is to be laid. They will be placed at a depth of 1.8 m below the ground surface. A relatively impermeable soil layer is found at a depth of 6.8 m below the surface. Hydraulic conductivity above this layer is 0.8 m/day. Average discharge of drainage system amounts to 2 mm/day. If an average water table depth of 1.20 m is to be maintained below the soil surface, what drain spacing should be adopted?

Solution 3.4

Refer **Figure 3.20b** for geometry.

r= 0.10 m

q = 0.002 m/day

K_1 = 0.8 m/day

h = 0.6 m

D_o = 6.8 – 1.8 = 5.0 m

D_1 = Do +1/2 h = 5+1/2 x 0.6 = 5.3m

Since soil is homogeneous Equation (3.93) and **Figure 3.23** can be used.

u = b + 2x2xr

= 0.25+4x0.10=0.65 m

$$h = 0.6 = q\frac{L^2}{8K_1D_1} + q\frac{L}{\pi K_1}\ln\frac{D_o}{u}$$

$$0.6 = 0.002X\frac{L^2}{8X0.8X5.30} + 0.002X\frac{L}{\pi X0.8}\ln\frac{5}{0.65}$$

and

$$L = \frac{-0.8 \pm \sqrt{0.64 + 4X0.03X300}}{2X0.03} = \frac{-0.8 \pm 6.05}{0.06}$$

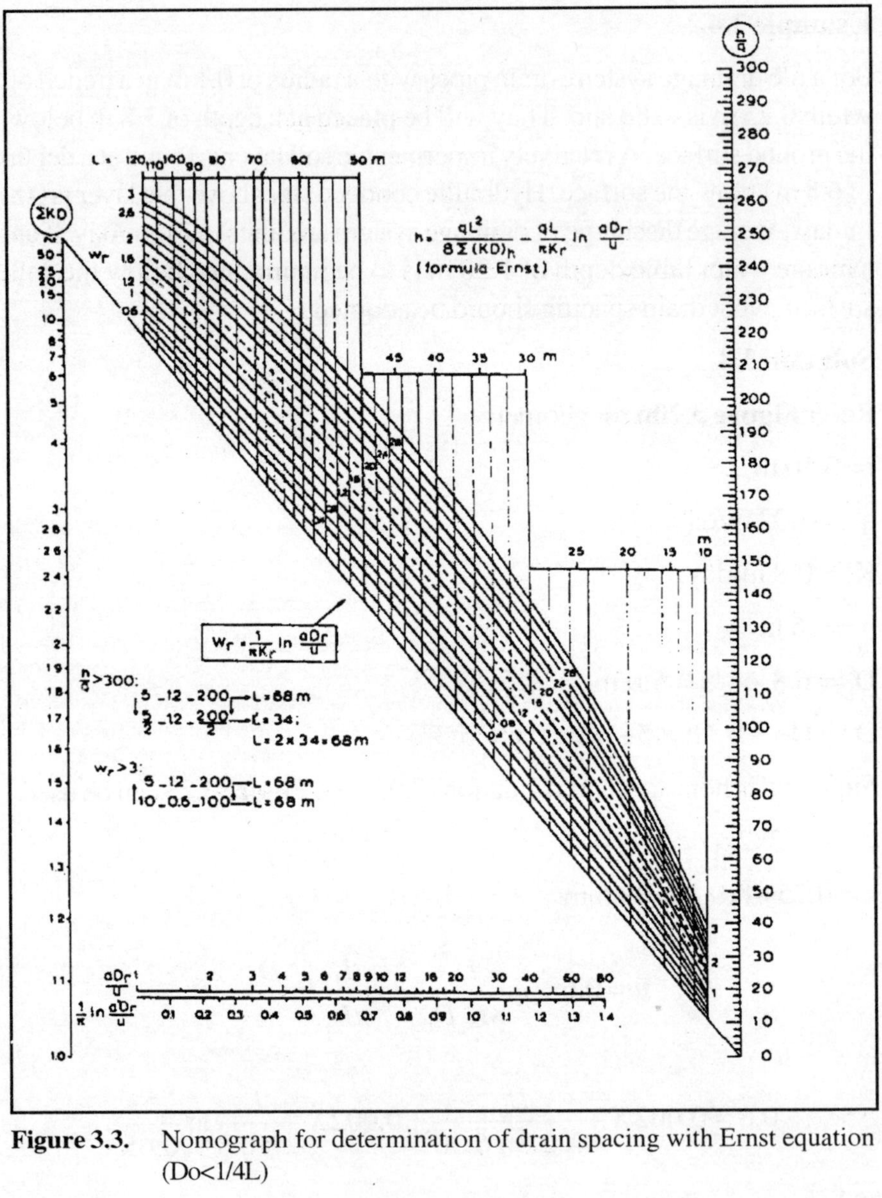

Figure 3.3. Nomograph for determination of drain spacing with Ernst equation (Do<1/4L)

L= 87.5 m

Since, L>0, L=87.5 m

The nomograph of **Figure 3.23** is used as,

Connect the point $\Sigma(KD)=K_1(D+1/2\ h)=0.8x5.30=4.2\ m^2/day$

On the left hand axis with the point h/q=0.6/0.002=300, on the right axis by a straight line. Intersecting with the curve.

$$W_r = \frac{1}{\pi K_r}\ln\frac{aD_r}{u} = \frac{1}{\pi X\, 0.8}\ln\frac{5}{0.65} = 0.8$$

Read on the vertical axis that L = 88 m.

Example 3.5

A soil consists of two distinct layers. For the upper layer K_1=0.2 m/day and for the lower layer K_2 = 2 m/day. The interface of two layers is at a depth of 0.5 m below the bottom of the drain ditch (**Figure 3.20d**). The thickness of the lower layer to an impermeable layer D_2 = 3 m. The ditch has the bottom width of 50 cm, side slope 1:1 and water depth y =30 cm. The hydraulic head is set at h = 1.20 m at a steady state discharge of q = 10 mm/day. Calculate the spacing between ditches.

Solution 3.5

h=1.2 m	Do=0.5 +0.3 = 0.8 m
q= 0.01 m/day	D_1= 0.8 +1/2x1.2 = 1.4 m
K_1= 0.2 m/day	D_2= 3.0 m
K_2= 2 m/day	u = 0.5 + 2x0.3√2 = 1.35 m
y = 0.3 m	

Step 1.

Assume Do<1/4 L so that Equation (3.92) is valid.

Step 2.

$$h_v = q\frac{D_v}{K_v} = q\frac{(h+y)}{K_1} = 0.01X\frac{1.2+0.3}{0.2} = 0.075m$$

$$h' = h - h_v = 1.2 - 0.075 = 1.125\ m$$

Step 3.

Since $K_2/K_1 = 10$ determine 'a' from **Figure 3.21.**

From point K_2/K_1=10 at the lower axis vertically upward to the line for D_2/D_O=3/0.8 = 3.8 (interpolate between 2 and 4) and read on the vertical axis a =4

$$\Sigma(KD)_h = K_1D_1 + K_2D_2 = 0.2 \times 1.4 + 2 \times 3.0 = 6.3 \text{ m}^2/\text{day}$$

$$W_r = \frac{1}{\pi K}\ln\frac{aD_r}{u} = \frac{1}{\pi K_1}\ln\frac{4D_o}{u} = \frac{1}{\pi X 0.2}\ln\frac{4X0.8}{1.35} = 1.37\,days/m$$

$$h' = 1.125m = q\frac{L^2}{8\Sigma(KD)_h} + q\frac{L}{\pi K_r}\ln\frac{aD_r}{u}$$

$$1.125 = 0.01X\frac{L^2}{8X6.3} + 0.01X1.37L$$

$$0.2\,L^2 + 13.7\,L - 1125 = 0$$

$$L = \frac{-13.7 + \sqrt{13.7^2 + 2X0.2X1125}}{2X0.2} = \frac{-13.7+33}{0.4}$$

L = 48 m

This value can also be found from **Figure 3.23.**

Since D_o = 0.8 m the condition D_o<1/4 L is fulfilled.

Example 3.6

Using data of **Example 3.5** except that D_o = 10 m calculate spacing between the ditches.

Solution 3.6

Step 1.

Since it is likely that Do will be more that ¼ L, the solution for a homogeneous soil, as given by Equation (3.93) will be applied. This means that second layer, whatever it's permeability or thickness, has no influence on the flow to the drains.

Following **Example 3.5**, step 2, the vertical hydraulic head = 0.075 m and h′ = 1.125 m.

Solving now Equation (3.93) for a=1, K_1D_1 = 0.2 X 10.6 = 2.1 m²/day, Do = 10 m and u=1.35 m, results in,

$$1.125 = \frac{0.01}{8X2.1}L^2 + \frac{0.01}{\pi X0.2}L\ln\frac{10}{1.35}$$

$$1.125 = 0.000595\, L^2 + (0.0159 \ln 7.41)\, L$$

$$L = 24 \text{ m}$$

Since D_o (=10 m) is of course more than ¼ L (6 m), so assumption $D_o >$ L is correct.

3.4.13 Kirkham Drain Spacing Equation

Kirkham (1958) gave an analytical solution for a problem similar to Hooghoudt viz. two dimensional flow, a regularly distributed rainfall over the area and drains not reaching impervious layer.

If R =rainfall rate (mm/day)

S = semi-spacing, L/2 (m)

ψ_o= quantity of water recharging a ditch from one side per unit time per unit length of ditch.

Then,

$$\psi_o = RS \qquad 3.100$$

3.4.13.1 Development of Stream and Potential Functions

ψ= stream function

φ = hydraulic head

Φ= velocity potential

$$\Phi = K\phi \qquad 3.101$$

Choose Φ such that $\phi = 0$ at y=0.

That is x-axis (as shown in **Figure 3.24**) will be the reference level for hydraulic head, and the distance z will accordingly be given by ϕ when ϕ is evaluated at x=x and y=0.

The height h will be given by ϕ when x=S and y=0.

The functions ψ and Φ, in order to be correct must satisfy the following differential equations.

$$\frac{\partial^2 \psi}{\partial x^2} + \frac{\partial^2 \psi}{\partial y^2} = 0 \qquad 3.102$$

$$\frac{\partial^2 \Phi}{\partial x^2} + \frac{\partial^2 \Phi}{\partial y^2} = 0 \qquad 3.103$$

and given boundary conditions.

At	y=0	$\psi = (x/S)\,\psi_o$	from x=0 to S Condition I
At	x=S	$\psi = \psi_o$	from y=0 to D Condition II
At	y=h	$\psi = \psi_o$	from x=0 to S Condition III
At	x=0	$\psi = \psi_o$	from y= b to D Condition IV
At	x=0	$\psi = [(y-a)/(b-a)]\psi_o$	from x=0 to S Condition V
At x=0		$\psi = 0$	from y=0 to a Condition VI

Let Am and Bm with m= 1,2,3,... be the arbitrary constants, then a function ψ as one may verify by differentiating and substituting into Equation (3.102), which satisfy (Equation 3.102 and) conditions II and III for any values of Am and Bm.

$$\psi = \sum_1^\infty Am.\sin\frac{m\pi x}{S}.\frac{\sinh[m\pi(D-y)/S]}{\sinh[m\pi D/S]} + \sum_1^\infty Bm.\sin\frac{m\pi(D-y)}{D}.\frac{\sinh[m\pi(S-x)/D]}{\sinh(m\pi S/D)} + \psi_o \qquad 3.104$$

Condition I and Equation (3.104) yield, after transposing the constant ψ_o to the left.

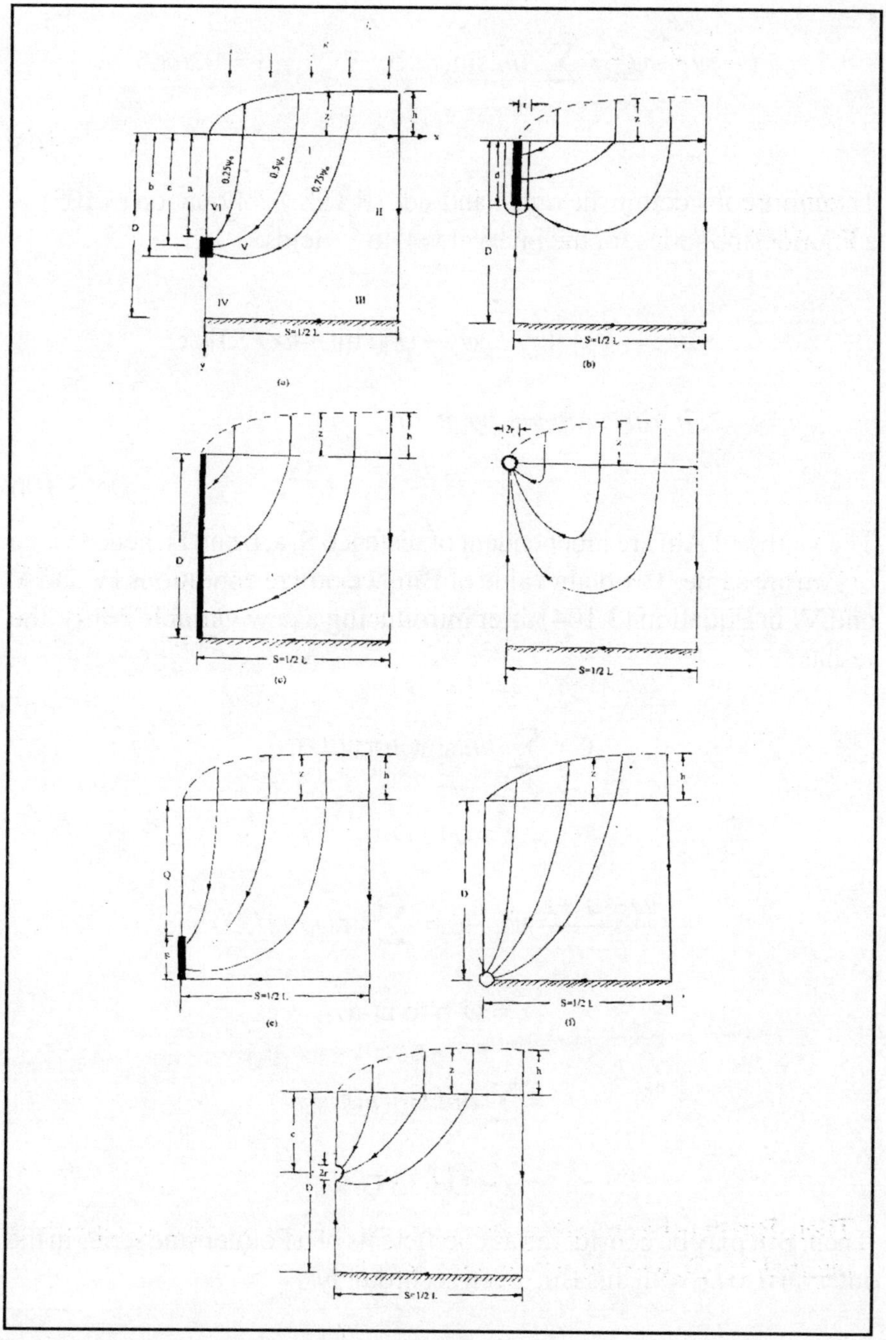

Figure 3.24. Semi-section of seepage region.

$$(\frac{x}{S})\psi_o - \psi_o = \sum_1^{\infty} Am.\sin(m\pi x / S),,,,,, x = 0..to..5$$

3.105

Therefore considering the right hand side (R.H.S.) of Equation (3.105) as a Fourier sine series for the interval x=0 to S yields.

$$Am = (\frac{2}{S})\int_0^S [(x/S)\psi_o - \psi][\sin(m\pi x / S)]dx$$

$$That`is`, Am = -2\psi_o \pi^{-1} m^{-1}$$

3.106

The values of Am are independent of distance S, a, b and D, hence value of Am are same. To obtain value of Bm, we utilize conditions IV and V and VI in Equation (3.104) after introducing a new variable z=h-y, the results

$$0 = \sum_1^{\infty} Bm.\sin(m\pi z / D),$$

z = 0 to D-b

$$(\frac{D-a-z}{b-a})\psi_o - \psi = \sum_1^{\infty} \sin(m\pi z / D)$$

z = D-b to D-a

$$-\psi_o = \sum_1^{\infty} Bm.\sin(m\pi z / D)$$

z = D-a to D

Then, Bm may be considered as coefficients of a Fourier sine series in the interval 0 to D, with this Bm given accordingly.

$$Bm = \frac{2}{D}\int_{D-b}^{D-a}[\frac{D-a-z}{b-a}\psi_o - \psi].\sin\frac{m\pi z}{D}dz + \frac{2}{D}\int_{D-a}^{D}(-\psi_o)\sin\frac{m\pi z}{D}dz$$

After performing the integration and simplification,

$$Bm = \frac{2\psi_o}{\pi}(-1)^m[\frac{1}{m} - \frac{D}{\pi m^2(b-a)}(\sin\frac{m\pi b}{D} - \sin\frac{m\pi a}{D})]$$

3.107

Our function ψ is thus completely determined by substituting the Am and Bm of Equation (3.106) and (3.107) into Equation (3.104).

To obtain the potential function Φ from ψ, we now utilize the Couchy-Riemann relation which are,

$$\frac{\partial\Phi}{\partial x} = \frac{\partial\psi}{\partial y}$$

3.108

$$\frac{\partial\Phi}{\partial y} = -\frac{\partial\psi}{\partial x}$$

3.109

The sought-for function Φ must satisfy (3.103), (3.108) and (3.109). By inspection of (3.103), (3.104) and (3.109) we see that Φ may be written as.

$$\Phi = \sum_1^\infty Am.\cos\frac{m\pi x}{S}.\frac{\cosh[m\pi(D-y)/S]}{\sinh(m\pi D/S)} + \sum_1^\infty Bm.\cos\frac{m\pi(D-y)}{D}.\frac{\cosh[m\pi(S-x)/D]}{\sinh[m\pi S/D]} + C$$

3.110

Where C is constant to be chosen such that Φ=Kϕ=0.

One may verify by differentiation of Equation (3.104) and (3.110) that (3.108) and (3.109) are satisfied by ψ and Φ of Equation (3.104) and (3.110).

A similar form of Equation (3.104) and (3.110) can be obtained by trigonometric reduction as.

$$\psi=\sum_{1}^{\infty}Am\sin\frac{m\pi x}{S}.\frac{\sinh[m\pi(D-y)/S]}{\sinh(m\pi D/S)}-\sum_{1}^{\infty}(-1)^{m}(Bm).\sin\frac{m\pi y}{D}.\frac{\sinh[m\pi(S-x)/D]}{\sinh[m\pi S/D]}+\psi_{o}$$

3.111

$$\Phi=\sum_{1}^{\infty}Am\cos\frac{m\pi x}{S}.\frac{\cosh[m\pi(D-y)/S]}{\sinh(m\pi D/S)}+\sum_{1}^{\infty}(-1)^{m}Bm\cos\frac{m\pi y}{D}.\frac{\cosh[m\pi(S-x)/D]}{\sinh[m\pi S/D]}+C$$

3.112

3.4.13.2 Rapidly converging series for ψ and Φ

The Am, as given by Equation (3.106) and Bm, as given by Equation (3.107), when substituted in Equation (3.111) and (3.112) yield ψ and Φ.

The results for y and F are rapidly converging infinite series, except for values of x and y near the sinks. Since we need to know F near the sink, we need to obtain rapidly converging series for RHS of Equation (3.112) for small x and y. We also need rapidly convergence series for RHS of Equation (3.111) for small x and y. We consider series of Equation (3.111).

ψ for **Figure 3.24a**

Observe the following two identities for the ratio of hyperbolic terms in Equation (3.111)

$$\frac{\sinh[m\pi(D-y)/S]}{\sinh(m\pi D/S)}=e^{-m\pi y/S}-\frac{e^{-m\pi S/D}.\sinh(m\pi y/S)}{\sinh(m\pi D/S)}$$

3.113

$$\frac{\sinh[m\pi(S-x)/D]}{\sinh(m\pi S/D}=e^{-m\pi S/D}-\frac{e^{-m\pi S/D}.\sinh(m\pi x/D)}{\sinh(m\pi S/D)}$$

3.114

Now substitute the RHS of Equation (3.106) and (3.107) and of (3.113) and (3.114) into (3.111) obtain, after some rearrangement,

$$\frac{\pi}{2}\frac{\psi_o-\psi}{\psi}=S_1-S_2+S_3-S_4-S_5+S_6$$

3.115

where,

$$S_1=\sum_{1}^{\infty}\frac{1}{m}\sin\frac{m\pi x}{S}e^{-m\pi y/S}$$

3.116

$$S_2=\sum_{1}^{\infty}\frac{1}{m}\sin\frac{m\pi x}{S}\frac{e^{-m\pi D/S}.\sinh(m\pi y/S)}{\sinh(m\pi D/S)}$$

3.117

$$S_3=\sum_{1}^{\infty}\frac{1}{m}\sin\frac{m\pi y}{D}e^{-m\pi x D/S}$$

3.118

$$S_4=\sum_{1}^{\infty}\frac{1}{m}\sin\frac{m\pi y}{D}\frac{e-m\pi S/D.\sinh(m\pi x/D)}{\sinh(m\pi S/D)}$$

3.119

$$S_5=\frac{D}{\pi(b-a)}\sum_{1}^{\infty}\frac{1}{m^2}(\sin\frac{m\pi b}{D}-\sin\frac{m\pi a}{D})\sin\frac{m\pi y}{D}e^{-m\pi x/D}$$

3.120

$$S_6=\frac{D}{\pi(b-a)}\sum_{1}^{\infty}\frac{1}{m^2}(\sin\frac{m\pi b}{D}-\sin\frac{m\pi a}{D}).\sin\frac{m\pi a}{D}\frac{e^{-m\pi S/D}.\sinh(m\pi x/D)}{\sinh(m\pi S/D)}$$

3.121

In Equation (3.115), S_1 and S_2 stem from term in Equation (3.111) is having coefficient Am. The remaining terms in Equation (3.115) having coefficient Bm.

The quantity S_2 as seen in the RHS of Equation (3.117), sums rapidly because of the factor exp (-mπD/S). Also, S_4 and S_6 sum extremely rapidly because of the factor exp (-mπS/D). The sums S_1 and S_3 converge very slowly for small x and y but this is immaterial. The sums S_1 and S_3 can be obtained in the closed form.

$$S_1 = \tan^{-1} \frac{\sin(\pi y / S)}{e^{\pi y / S} - \cos(\pi x / S)} \tag{3.122}$$

$$S_3 = \tan^{-1} \frac{\sin(\pi y / D)}{e^{\pi x / D} - \cos(\pi y / D)} \tag{3.123}$$

The remaining sum to consider, S_5 of Equation (3.120) does not converge rapidly for small x and y. But neither does it converge very slowly. A very rapidly converging series for RHS of Equation (3.120) can be obtained. If one sums the first 100 terms of RHS of Equation (3.120) as it stands the absolute value of error, committed by omitting the remaining terms, will be less than.

$$(\sum_{100}^{\infty} \frac{1}{m^2}) \frac{D}{\pi(b-a)} e^{-101\pi x / D} = 0.005000 \frac{D}{\pi(b-a)} e^{-101\pi x / D} \tag{3.124}$$

And hence negligible even if x=0.

Φ for Figure 3.24a

Observe that the ratio of the hyperbolic terms in Equation (3.112) may be written as.

$$\frac{\cosh[m\pi(D-y)/S]}{\sinh(m\pi D / x)} = e^{-m\pi y / S} + \frac{e^{-m\pi D / S} \cosh(m\pi y / S)}{\sinh(m\pi D / S)} \tag{3.125}$$

$$\frac{\cosh[m\pi(S-x)/D]}{\sinh(m\pi S/D)} = e^{-m\pi x/D} + \frac{e^{-m\pi DS/D}\cosh(m\pi x/D)}{\sinh(m\pi S/D)}$$

3.126

Now substitute the value of RHS of Equation (3.106) and (3.107) and of (3.125) and (3.126) into (3.112) to obtain, after rearranging.

$$\frac{\pi}{2} = \frac{\Phi - C}{\psi_o} = V_{(x,y)}$$

3.127

where,

$$V(x,y) = -T_1 - T_2 + T_3 + T_4 - T_5 - T_6$$

1.128

$$T_1 = \sum_1^{\infty} \frac{1}{m} \cos \frac{m\pi x}{S} e^{-m\pi y/S}$$

3.129

$$T_2 = \sum_1^{\infty} \frac{1}{m} \cos \frac{m\pi x}{S} \frac{e^{-m\pi D/S}.\cosh(m\pi y/S)}{\sinh(m\pi D/S)}$$

3.130

$$T_3 = \sum_1^{\infty} \frac{1}{m} \cos \frac{m\pi y}{D} e - m\pi x/D$$

3.131

$$T_4 = \sum_1^{\infty} \frac{1}{m} \cos \frac{m\pi y}{D} . \frac{e^{-m\pi S/D}\cosh(m\pi x/D)}{\sinh(m\pi S/D)}$$

3.132

$$T_5 = \frac{D}{\pi(b-a)} \sum_1^{\infty} \frac{1}{m^2} (\sin \frac{m\pi b}{D} - \sin \frac{m\pi a}{D}).\cos \frac{m\pi y}{D} e^{-m\pi x/D}$$

3.133

$$T_6 = \frac{D}{\pi(b-a)} \sum_1^{\infty} \frac{1}{m^2} (\sin \frac{m\pi b}{D} - \sin \frac{m\pi a}{D}).\cos \frac{m\pi y}{D} \frac{e^{-m\pi S/D}.\cosh(m\pi x/D)}{\sinh(m\pi S/D)}$$

3.134

In Equation (3.129) to (3.134) the sums T_1 and T_3 are very slowly convergent for small x and y but this does not matter.

Instead of Equation (3.119) and (3.131) we may use,

$$T_1 = -\frac{1}{2} \ln[2e^{-\pi y/S} (\cosh \frac{\pi y}{S} - \cos \frac{\pi x}{S})]$$

3.135

$$T_3 = -\frac{1}{2} \ln[2e^{-\pi x/D} (\cosh \frac{\pi x}{D} - \cos \frac{\pi y}{D})]$$

3.136

In the other sums in Equation (3.129) to (3.134) converges very rapidly.

ψ and Φ for Figure 3.24b

For Figure 3.24b put a=0, b=d in Equation (3.115) and (3.128) and auxillary equation. The equation will remain same except (3.120) and (3.121), and (3.133) and (3.134) which become.

$$S_5^b = \frac{D}{\pi d} \sum_1^{\infty} \frac{1}{m^2} \sin \frac{m\pi d}{D} \sin \frac{m\pi y}{D} e^{-m\pi x/D}$$

3.137

$$S_6^b = \frac{D}{\pi d} \sum_1^{\infty} \frac{1}{m^2} \sin \frac{m\pi d}{D} \sin \frac{m\pi y}{D} . \frac{e^{-m\pi S/D}.\sinh(m\pi x/D)}{\sinh(m\pi S/D)}$$

3.138

$$T_5^b = \frac{D}{\pi d}\sum_1^{\infty}\frac{1}{m^2}\sin\frac{m\pi d}{D}\cos\frac{m\pi y}{D}e^{-m\pi x/D}$$

3.139

$$T_6^b = \frac{D}{\pi d}\sum_1^{\infty}\frac{1}{m^2}\sin\frac{m\pi d}{D}\cos\frac{m\pi y}{D}.\frac{e^{-m\pi S/D}.\cosh(m\pi x/D)}{\sinh(m\pi S/D)}$$

3.140

where, T_6^b means T_6 for the case of **Figure 3.24b.** Therefore (3.115) becomes,

$$\frac{\pi}{2}(\frac{\psi_o - \psi^b}{\psi_o}) = S_1 - S_2 + S_3 - S_4 - S_5^b + S_6^b$$

3.141

and Equation (3.127) and (3.128) become,

$$V_{(x,y)}^b = \frac{\pi}{2}(\frac{\Phi - C^b}{\psi_o}) = -T_1 - T_2 + T_3 + T_4 - T_5^b - T_6^b$$

3.142

ψ and Φ for Figure 3.24c

Put d=D in Equation (3.137) to (3.140) and obtain,

Since sin mπ =0

$$S_5^C = 0, S_6^C = 0, T_5^C = 0, T_6^C = 0$$

3.143

Therefore Equation (3.115) and (3.128) become.

$$\frac{\pi}{2}\frac{\psi_o - \psi^C}{\psi_o} = S_1 - S_2 + S_3 - S_4$$

3.144

$$V^{C}_{(x,y)} = \frac{\pi}{2}\frac{\Phi_C - C^c}{\psi_o} = -T_1 - T_2 + T_3 + T_4$$

3.145

ψ and Φ for Figure 3.24d

Let in Equation (3.120) and (3.121), a=0, b=d and then d→0, so that (width d dropping out of result as should)

$$\frac{h}{\pi(b-a)}\frac{1}{m^2}(\sin\frac{m\pi b}{D} - \sin\frac{m\pi a}{D}) = \frac{D}{\pi d}.\frac{1}{m^2}\sin\frac{m\pi d}{D} \rightarrow \frac{D}{\pi d}\frac{1}{m^2}\frac{m\pi d}{D} = \frac{1}{m}$$

3.146

Therefore using Equation (3.146) in (3.120) and (3.121) and comparing with Equation (3.118) and (3.119), find

$$S_6^d = S_4, \quad S_5^d = S_3$$

3.147

So that (3.115) becomes

$$\frac{\pi}{2}\frac{\psi_o - \psi^d}{\psi_o} = S_1 - S_2$$

3.148

Similarly,

$$T_6^d = T_4, \quad T_5^d = T^3$$

3.149

So that Equation (3.128) becomes

$$V^{d}_{(x,y)} = \frac{\pi}{2}\frac{\Phi^d - C^d}{\psi_o} = -T_1 - T_2$$

3.150

ψ and Φ for Figure 3.24e

Put in Equation (3.120) and (3.121) b=D and a=D-g where g = height of line sink.

$$S_5^e = \frac{D}{\pi g}\sum_1^{\infty}(-1)^m \frac{1}{m^2}\sin\frac{m\pi g}{D}\sin\frac{m\pi y}{D}e^{-m\pi x/D}$$

3.151

$$S_6^e = \frac{D}{\pi g}\sum_1^{\infty}(-1)^m \frac{1}{m^2}\sin\frac{m\pi g}{D}\sin\frac{m\pi y}{D}.\frac{e^{-m\pi S/D}.\sinh(m\pi x/D)}{\sinh(m\pi S/D)}$$

3.152

Whence, Equation (3.115) becomes,

$$\frac{\pi}{2}\frac{\psi_o - \psi^e}{\psi_0} = S_1 - S_2 + S_3 - S_4 - S_5^e + S_6^e$$

3.153

Similarly from Equation (3.133) and (3.134)

$$T_5^e = \frac{D}{\pi g}\sum_1^{\infty}(-1)^m \frac{1}{m^2}\sin\frac{m\pi g}{D}.\cos\frac{m\pi y}{D}e^{-m\pi x/D}$$

3.154

$$T_6^e = \frac{D}{\pi g}\sum_1^{\infty}(-1)^m \frac{1}{m^2}\sin\frac{m\pi g}{D}\cos\frac{m\pi y}{D}\frac{e^{-m\pi S/D}.\cosh(m\pi x/D)}{\sinh(m\pi S/D)}$$

3.155

and Equation (3.128)

$$V_6^e = \frac{\pi}{2}\frac{\Phi_e - e^e}{\psi_o} = -T_1 - T_2 + T_3 + T_4 - T_5 - T_6$$

3.156

ψ and Φ for Figure 3.24f

In Equation (3.151), (3.152), (3.154) and (3.156), let g→0 so that sin mπg/D→mπg/D and find.

$$S_5^f = \sum_1^\infty (-1)^m \frac{1}{m} \sin \frac{m\pi y}{D} e^{-m\pi x/D} \tag{3.157}$$

$$S_6^f = \sum_1^\infty (-1)^m \frac{1}{m} \sin \frac{m\pi y}{D} \frac{e^{-m\pi S/D}.\sinh(m\pi x/D)}{\sinh(m\pi S/D)} \tag{3.158}$$

$$T_5^f = \sum_1^\infty (-1)^m \frac{1}{m} \cos \frac{m\pi y}{D} e^{-m\pi x/D} \tag{3.159}$$

$$T_6^f = \sum_1^\infty (-1)^m \frac{1}{m} \cos \frac{m\pi y}{D} \frac{e^{-m\pi S/D} \cosh(m\pi x/D)}{\sinh(m\pi S/D)} \tag{3.160}$$

Therefore Equation (3.115) and (3.128) become

$$\frac{\pi}{2} \frac{\psi_o - \psi^f}{\psi_o} = S_1 - S_2 + S_3 - S_4 - S_5^f + S_8^f \tag{3.161}$$

$$V^f = \frac{\pi}{2} \frac{\Phi^f - C^f}{\psi_o} = -T_1 - T_2 + T_3 + T_4 - T_5^f - T_6 \tag{3.162}$$

Sums in Equation (3.157) and (3.159) converge very slowly for small x/D. This does not matter since they can be summed analytically.

$$S_5^f = -\tan^{-1}\frac{\sin(\pi y/D)}{e^{-\pi x/D}-\cos(\pi y/D)}+\tan^{-1}\frac{\sin(2\pi y/D)}{e^{-2\pi x/D}-\cos(2\pi y/D)}$$

3.163

$$T_5^f = \frac{1}{2}\ln[2e^{-\pi x/D}(\cosh\frac{\pi x}{D}-\cos\frac{\pi y}{D})]-\frac{1}{2}\ln[2e^{-2\pi x/D}(\cosh\frac{2\pi x}{D}-\cos\frac{2\pi y}{D})]$$

3.164

We find for ($T_3+T_5^f$), in view of Equation (3.137) and (3.105)

$$T_3+T_5^f = -\frac{1}{2}[\ln 2-\frac{2\pi x}{D}+\ln(\cosh\frac{2\pi x}{D}-\cos\frac{2\pi y}{D})]$$

3.165

ψ and Φ for Figure 3.24g

Denote the semi-height (b-a)/2 of the strip by Δ. That is b-a =2Δ, b= c+Δ, a=c-Δ, find upper letting Δ→0.

$$\frac{D}{\pi(b-a)}(\sin\frac{m\pi b}{D}-\sin\frac{m\pi a}{D})=\frac{D}{2\pi\Delta}[\sin\frac{m\pi(c+\Delta)}{D}-\sin\frac{(c-\Delta)}{\Delta}]$$

$$=\frac{D}{2\pi\Delta}.2\sin\frac{m\pi\Delta}{D}\cos\frac{m\pi c}{D}\rightarrow m\cos\frac{m\pi c}{D}$$

So that (dropping out Δ),

$$S_6^g = \sum_1^\infty\frac{1}{m}\cos\frac{m\pi c}{D}\sin\frac{m\pi y}{D}\frac{e^{-m\pi S/D}.\sinh(m\pi x/D)}{\sinh(m\pi S/D)}$$

3.166

$$S_6^g = \sum_1^\infty\frac{1}{m}\cos\frac{m\pi c}{D}\sin\frac{m\pi y}{D}\frac{e^{-m\pi S/D}.\sinh(m\pi x/D)}{\sinh(m\pi S/D)}$$

3.167

$$T_5^g = \sum_1^{\infty} \frac{1}{m} \cos\frac{m\pi c}{D} \cos\frac{m\pi y}{D} e^{-m\pi x/D}$$

3.168

$$T_6^g = \sum_1^{\infty} \frac{1}{m} \cos\frac{m\pi c}{D} \cos\frac{m\pi y}{D} \frac{e^{-m\pi S/D} .\cosh(m\pi x/D)}{\sinh(m\pi S/D)}$$

3.169

Accordingly Equation (3.115) and (3.128) become,

$$\frac{\pi}{2} \frac{\psi_o - \psi^g}{\psi_o} = S_1 - S_2 + S_3 - S_4 - S_5^g + S_8^g$$

3.170

$$V_{(x,y)}^g = \frac{\pi}{2} \frac{\Phi^g - C^g}{\psi_o} = -T_1 - T_2 + T_3 + T_4 - T_5^g - T_6^g$$

3.171

The sums in Equation (3.166) and (3.168) converges slowly for small x and y. Therefore for small x and y,

$$S_5^g = \frac{1}{2}\tan^{-1}\frac{\sin[\pi(y+c)/D]}{e^{\pi x/D} - \cos[\pi(y+c)/D]} + \frac{1}{2}\tan^{-1}\frac{\sin[\pi(y-c)/D]}{e^{\pi x/D} - \cos[\pi(y-c)/D]}$$

3.172

$$T_6^g = -\frac{1}{4}\ln\{2e - \pi x/D[\cosh\frac{\pi x}{D} - \cos\frac{\pi(y+c)}{D}]\} - \frac{1}{4}\ln\{2e^{-\pi x/D}[\cosh\frac{\pi x}{D} - \cos\frac{\pi(y-c)}{D}]\}$$

3.173

3.4.13.3 Evaluation of Constant 'C'

In considering the constant C the situation for **Figure 3.24 a, d, f** and **g** considered separately.

For **Figure 3.24d**, let Φ(xo,0) be the value of an equipotential surface passing through the point (r,o), and let us identify the equipotential passing through (r,0) with the shape of a drainage facility assumed running full to the level y=0, that is r is the semi-width of the ditch or tube. Then since it has been agreed to take the reference level at the x-axis, we see that the left hand equality of Equation (3.142) becomes,

$$V^{b}_{(x,y)} = \frac{\pi}{2}\frac{\Phi - C^{b}}{\psi_{o}} = -T_1 - T_2 + T_3 + T_4 - T_5^{B} - T_6^{B}$$

$$\frac{\pi}{2}\frac{0 - C_{b}}{\psi_{o}} = V^{b}_{(r,0)}$$

$$C_{b} = -(\frac{2}{\pi})\psi_{o}V^{b}_{(r,0)}$$

3.174

Where $V^{b}_{(r,0)}$ is to be obtained from RHS of Equation (3.142) by putting x and y in the T^{s} respectively equal to r and o. Similarly one may obtain from Equation (3.145) and (3.150).

$$C^{c} = -(\frac{2}{\pi})\psi_{o}V^{C}_{(r,0)}$$

3.175

$$C^{d} = -(\frac{2}{\pi})\psi_{o}V^{d}_{(r,0)}$$

3.176

where,

$V^{c}_{(r,0)}$ is evaluated from RHS of Equation (3.145) and $V^{d}_{(r,0)}$ from RHS of Equation (3.150).

To obtain C for **Figure 3.24 a, e, f** and **g** we observe that the water table stands above the sink at all points, including point immediately over the sink. Therefore, we can not, as before, simultaneously identify a point on the sink with a point as the water table. Instead we must take, in Equation (3.126), $V_{(x,y)}=V_{(0,0)}$ with Φ=0, we obtain then for cases a, e, f and g using a composite notation to show four equations in one.

$$C^{aefg} = -(\frac{2}{\pi})\psi_o V^{aefg}_{(0,0)}$$

3.177

where,

V^a is obtained from Equation (3.128), (3.135), (3.130), (3.136), (3.132), (3.130) and (3.134); $V_{(o,o)}$ from (3.145), (3.135), (3.130), (3.136), and (3.132), $V^f(0,0)$ from (3.162), (3.135), (3.130), (3.136), and (3.132) and $V^g(0,0)$ from (3.171), (3.135), (3.130), (3.136), (3.132), (3.172), (3.169), (3.164), and (3.160).

In Equations (3.128), (3.145), (3.162), and (3.171) notice that $(-T_1+T_3)$ in view of (3.135) and (3.136) simplifies to

$$-T_1 + T_3 = \frac{1}{2}\ln\frac{e^{-\pi y/S}[\cosh(\pi y/S) - \cos(\pi x/S)]}{e^{-\pi x/D}[\cosh(\pi x/D) - \cos(\pi y/D)]}$$

On expression which when x and y both approach to zero reduces to the form,

$$-T_1 + T_3 = \ln\frac{D}{S} + \frac{\pi}{2}(\frac{x}{D} - \frac{y}{S}), \ldots\ldots x, y \to 0$$

In particular when x and y are both zero as for $V_{(0,0)}$ we have

$$-T_1 + T_3 = \ln\frac{D}{S}, \ldots.(x, y \to 0)$$

3.4.13.4 Formula for the hydraulic head

Using the values of C^{bcd} of Equation (3.174), (3.175) and (3.178) in the equalties of the left of Equation (3.142), (3.145) and (3.150) and using then Equation (3.100) and (3.101) we find.

$$\phi^{bcd}_{(x,y)} = \frac{2SR}{K}.\frac{1}{\pi}[V^{bcd}_{(x,y)} - V^{bcd}_{(r,0)}]$$

3.178

And similarly from Equation (3.177) and (3.127) and further from Equation (3.156), (3.162), (3.171), (3.100) and (3.101) we find.

$$\phi^{aefg}_{(x,y)} = \frac{2SR}{K}.\frac{1}{\pi}[V^{aefg}_{(x,y)} - V^{aefg}_{(r,0)}]$$

3.179

The quantity ϕ^{bcd} in Equation (3.178) gives the hydraulic head in the soil for cases b, c and d of **Figure 3.24.**

3.4.13.5 Formula in submerged drain tube

The hydraulic head at all points on a nearly circular drain tube of **Figure 3.24g** whose circumference includes a point (x,y)=(r,c) is obtained from expression (3.179) with the result.

$$\phi^{g}_{(r,c)} = \frac{2SR}{K}.\frac{1}{\pi}[V^{g}_{(r,c)} - V^{g}_{(0,0)}]$$

3.180

Similarly for **Figure 3.24f**

$$\phi_{(r,D)} = \frac{2SR}{K}.\frac{1}{\pi}[V^{f}_{(r,D)} - V^{f}_{(0,0)}]$$

3.181

We point out an important matter in applying Equation (3.180) and (3.181) that unless in Equation (3.180) the value of r is such that ϕ=-C, the drain tube will have to be under suction.

3.4.13.6 Height of water table

$$\phi^{bcd}_{(x,0)} = \frac{2SR}{K}.\frac{1}{\pi}[V^{bcd}_{(x,0)} - V^{bcd}_{(r,0)}]$$

3.182

Above equation was obtained by putting y=0 in Equation (3.178) to obtain height $\phi^{bcd}_{(x,o)}$ (=z of **Figure b, c, d**) of the water table above reference level. Similarly, from Equation (3.179).

$$\phi^{aefg}_{(x,0)} = \frac{2SR}{K} \cdot \frac{1}{\pi}[V^{aefg}_{(x,0)} - V^{aefg}_{(0,0)}]$$

3.183

Putting x=S in Equation (3.182) and (3.183) gives for their left hand sides the height h of the cases in Figure a, b, c, d, e, f and g.

Case d is of particular interest. For it we have from Equation (3.182) [with z^d used for $\phi^d_{(x,0)}$]

$$z^d = \frac{2SR}{K} \cdot \frac{1}{\pi}[V^d_{(x,0)} - V^d_{(r,0)}]$$

3.184

Or in elemental terms, using Equation (3.150), (3.135) and (3.130) with (3.184) and simplifying.

$$z^d = \frac{2S}{R}\frac{1}{\pi}[\ln\frac{\sin(\pi x/2S)}{\sin(\pi r/2S)} + \sum_{m=1}^{\infty}\frac{1}{m}(\cos\frac{m\pi r}{S} - \cos\frac{m\pi x}{S})\frac{e^{-m\pi D/S}}{\sinh(m\pi D/S)}]$$

3.185

In numerical computation one should use, to save work the identity.

$$\frac{e^{-m\pi D/S}}{\sinh(m\pi D/S)} = \coth(m\pi D/S) - 1$$

3.186

The maximum height h of water table is obtained by putting x=S in Equation (3.185). The result is, utilizing (3.186) and noting the practical work that sin (1/2 πr/S).

$$h = \frac{2SR}{K}\frac{1}{\pi}[\ln\frac{2S}{\pi r} + \sum_{m=1}^{\infty}\frac{1}{m}(\cos\frac{m\pi r}{S} - \cos m\pi)\coth\frac{m\pi D}{S} - 1)]$$

3.187

When h/S→0 in Equation (3.187) (If impermeable layer is at great depth). The Equation (3.187) reduces to,

$$h = \frac{2SR}{K}\frac{1}{\pi}\ln\frac{2S}{\pi r}$$

3.188

Equation (3.187) can also be rewritten as, by putting S=L/2 and R=q.

$$h = \frac{qL}{K}.\frac{1}{\pi}[\ln\frac{L}{\pi r} + \sum_{m=1}^{\infty}\frac{1}{m}(\cos\frac{2m\pi r}{L} - \cos(m\pi)).\coth\frac{2m\pi D}{L} - 1]$$

3.189

In Equation (3.189) let F_K be defined by,

$$F_K = \frac{1}{\pi}[\ln\frac{L}{\pi r} + \sum_{m=1}^{\infty}(\cos\frac{2m\pi r}{L} - \cos(m\pi)).\coth\frac{2m\pi D}{L} - 1)]$$

3.190

Then Equation (3.189) becomes,

$$h = \frac{qL}{K}F_K$$

3.191

Toksoz and Kirkham (1961) presented a tabular form of F_K values as shown in **Table 3.3**. It is found that the Fk values of Kirkham are very close to the F_H values of Hooghoudt, so that both the Hooghoudt and Kirkham equations give almost identical results **(Wesseling, 1964)**.

A graphical solution of Kirkham's equation is presented in **Figure 3.25** (modified after **Toksöz and Kirkham, 1961**).

In the solution represented by Equation (3.191) the flow in the upper region has been neglected. In a later paper Kirkham (1960) reported that if vertical flow is assumed in the upper region, the hydraulic head should be multiplied by $(1\text{-}q/K)^{-1}$. Since this term relates the flow in the layer above drain level, the general equation for a two layered problem is (Wesseling, 1964),

Table 3.3 Values of F_K according to Toksöz and Kirkham.

L/D	100	50	25	12.5	6.25	3.125	1.5625	0.78125
D/2r	-	-	-	-	-	-		
8192	-	-	-	-	-	-	-	2.654
4096	-	-	-	-	-	-	2.65	2.43
2048	-	-	-	-	-	2.66	2.43	2.21
1024	-	-	-	-	2.84	2.45	2.21	1.99
512	-	-	-	3.40	2.63	2.23	1.99	1.76
256	-	-	4.76	3.19	2.40	2.01	1.76	1.54
128	-	7.64	4.53	2.96	2.19	1.78	1.54	1.32
64	13.67	7.43	4.31	2.74	1.96	1.57	1.32	1.10
	13.47	7.21	4.01	2.52	1.74	1.35	1.10	0.88
16	13.27	6.99	3.86	2.30	1.52	1.13	0.88	0.66
8	13.02	6.76	3.64	2.08	1.30	0.90	0.66	0.44
4	12.79	6.54	3.42	1.86	1.08	0.68	0.44	-
2	12.57	6.32	3.20	1.63	0.85	0.46	-	-
1	12.33	6.08	2.95	1.40	0.62	-	-	-
0.5	12.03	5.77	2.66	1.11	-	-	-	-
0.25	11.25	5.29	2.20	-	-	-	-	-

$$h = \frac{qL}{K_b} \cdot \frac{1}{(1-\frac{q}{K_a})} F_K \qquad 3.192$$

where,

K_a = hydraulic conductivity of the upper soil layer above the drain level, [L/T] and

K_b = hydraulic conductivity of the lower soil layer below the drain level, [L/T]

The interface in this case is assumed to be at drainage level.

3.4.14 Application of Kirkham drain spacing equation

Problem 3.7: Following information were collected for the installation of subsurface drainage in sub-humid region of Nainital Tarai region of U.P.

1. Radius of the drain pipe available in the market = 0.10 m
2. Depth of impervious layer below drainage level = 5.0 m
3. Steady drain discharge from a water balance study of the area = 0.002 m/day
4. Height of water table above drainage level to be maintained in the area = 0.6 m
5. Average value of saturated hydraulic conductivity = 0.8 m/day.

Work out drain spacing for the area to maintain water table at desired depth.

Solution 3.7:

i) Use of drain spacing formula

a) Calculate D/2r = 5/(2x0.1) = 25

b) Assume L = 60 m

c) Calculate L/D = 60/5 = 12

d) Find out the value of F_k from **Table 3.3** which is, F_k=2.36

e) Calculate drain spacing as,

$L = Kh/qF_K = (0.8x0.6)/(0.002x2.96) \approx 102$ m

Since the calculated value of L and assumed value of L are not the same, assume another value of L and repeat the calculation as below.

a) Assume L= 84 m.

a) Calculate L/D=84/5= 16.8

b) Find out the value of F_K= 2.96 from **Table 3.3**

c) Calculate drain spacing as,

$$L = Kh/qF_K = (0.8x0.6)/(0.002x2.96) = 81 \text{ m}$$

Since the assumed value of L and calculated value of L are almost identical, therefore L = 84 m can be adopted finally.

ii) Use of nomograph for drain spacing

Calculation of drain spacing may be also done using the Nomograph (**Figure 3.25**). Drain spacing can be worked out for the above data set as follows.

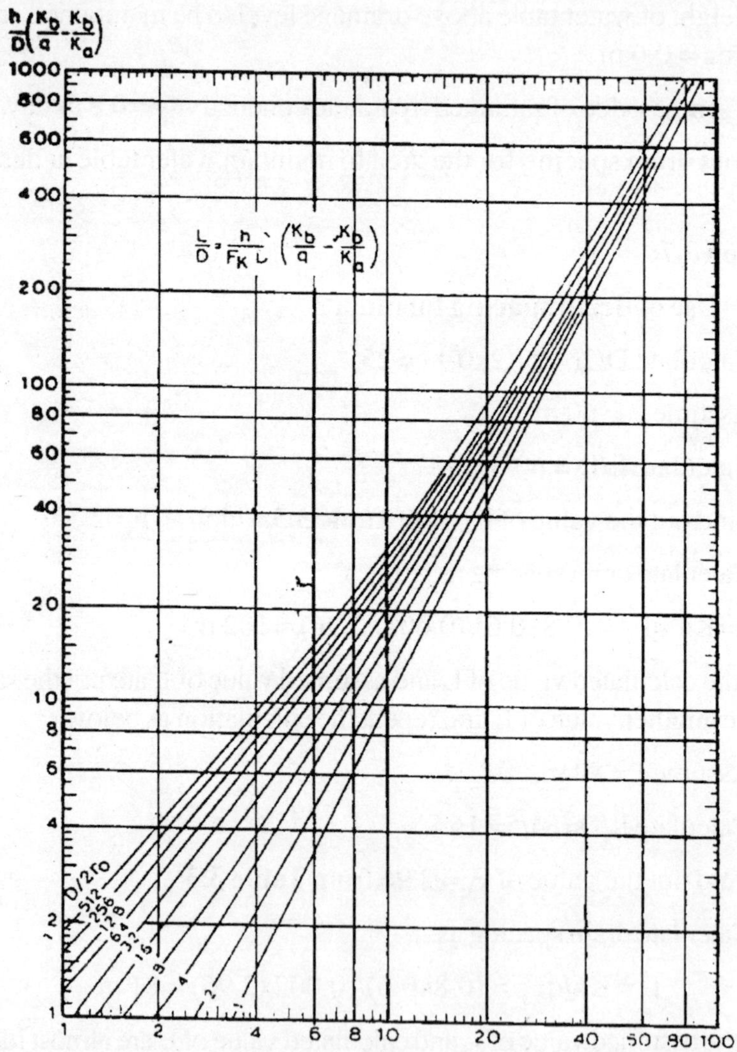

Figure 3.25. Nomograph for determination of spacing between the drains (modified after **Toköz and Kirkham,** 1961)

a) Calculate the ratio of h/D= (K_b/q –K_b/K_a)=0.6/5(0.8/0.002 – 0.8/0.8)= 48.

b) Move on vertical axis and locate the value h/D on the axis.

c) Move horizontally form this point till the line marked D/2r = 5/(2x0.1) = 25, which is found by interpolation between the lines marked 16 and 32.

d) Move from this point vertically downward and read the value of L/D= 15 on x-axis.

e) With D=5 m, calculate L = 5x17 = 85 m.

3.4.15 Dagan drain spacing equation

Analogous to Hooghoudt, **Dagan (1964)** thought the flow to be composed of a radial flow in the area between the drain and a distance ½ D√2 away from the drain; and mainly horizontal, flow in the area between ½ D√2 plane and the mid plane between the drains.

3.4.15.1 Flow to a sink in a confined layer

The flow net for this case is shown in **Figure 3.26**. The complex velocity can be obtained by the image method.

$$W_1 = u_1 + iv_1 = -\frac{Q}{2D}\frac{\sinh\frac{\pi z}{D}}{\cosh\frac{\pi z}{D} - \cos\frac{\pi y_o}{D}} = -\frac{Q}{2D}\frac{\sinh(\frac{\pi x}{D} + i\frac{\pi y}{D})}{\cosh(\frac{\pi x}{D} + i\frac{\pi y}{D}) - \cos\frac{\pi y_o}{D}} \tag{3.193}$$

where,

D = depth of aquifer at x=0, [L]

Q = sink discharge, [L^2/T]

$$z = x + iy \tag{3.194}$$

a complex variable.

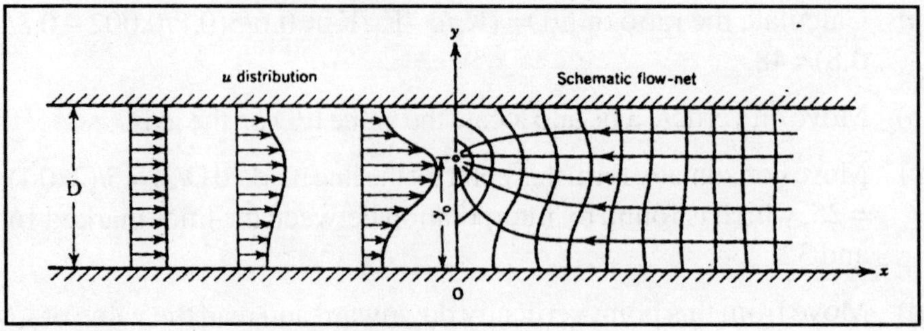

Figure 3.26. Confined flow towards drain (point sink).

From Equation (3.193), the expression for the horizontal component can be derived by separation of real parts.

$$W_1 = -\frac{Q}{2D} \frac{(\cosh\frac{\pi x}{D} - \cos\frac{\pi y}{D}\cos\frac{\pi y_o}{D})\sinh\frac{\pi x}{D}}{(\cosh\frac{\pi z}{D}\cos\frac{\pi y}{D} - \cos\frac{\pi y_o}{D})^2 + \sinh^2\frac{\pi x}{D}\sin\frac{\pi y}{D}}$$

3.195

At a distance varying from x = D (for yo =D/2) and x=2D (for yo=0, or yo=D), the distribution of u along the vertical is practically uniform. Maximum deviation of u from:

$$u_{av} = \frac{Q}{2D}$$

3.196

resulting from Equation (3.195) is 1% at these distances from sink. The complex potential is expressed by:

$$f_1(z) = \phi_1(x,y) + i\psi_1(x,y) = \frac{Q}{2}\ln(\cosh\frac{\pi z}{D} - \cos\frac{\pi y_o}{D}) + constant$$

3.197

where,

$f_{(z)}$ = complex potential = $\phi + i\psi$

u, v = horizontal and vertical component of velocity vector, [L/T].

y_o = elevation of the tile drain axis, [L]

$\phi_{(x,y)}$ = K(p/r + y) = K ϕ = velocity potential

$\psi_{(x,y)}$ = stream function

Separating the real part $\phi_{1(x,y)}$ and expressing its value on the upper boundary y = D yields.

$$\phi_1(x,D) = \frac{Q}{2\pi}\ln(\cosh\frac{\pi x}{D} + \cos\frac{\pi y_0}{D}) + cons\tan t$$

3.198

3.4.15.2 Potential of continuous distribution of sink

Flow net for this case is shown in **Figure 3.27.** The complex potential which results from summation of the elementary potential of uniformly distributed sinks of equal intensity, qdy, along D-l≤y≤D is,

$$f_2(z) = \phi_2(x,y) + i\psi_2(x,y) = \int_{D-l}^{D} \frac{q}{2\pi}\ln(\cosh\frac{\pi z}{D} - \cos\frac{\pi\xi}{D})d\xi + cons\tan t$$

$$= \frac{q}{2\pi}\int_0^l \ln(\cosh\frac{\pi z}{D} + \cos\frac{\pi\xi}{D})d\xi + cons\tan t$$

3.199

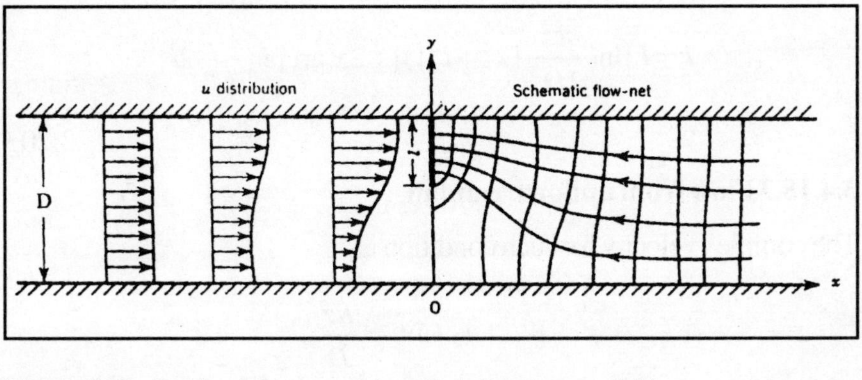

Figure 3.27. Confined flow towards a ditch (continuous sink)

where,

q = specific ditch discharge, [L/T]

ξ = integration variable

The potential variation on the upper boundary is given by.

$$\phi_2(x, D) = \frac{q}{2\pi}\int_0^l \ln(\cosh\frac{\pi x}{D} - \cos\frac{\pi\xi}{D})d\xi + cons\tan t \tag{3.200}$$

The integral,

$$I = \int_0^l \ln(\cosh\frac{\pi x}{D} - \cos\frac{\pi\xi}{D})d\xi \tag{3.201}$$

may be performed for a small depth (l<0.05H) by developing cos πξ/D, into a series.

$$\cos\frac{\pi\xi}{D} = 1 - \frac{(\pi\xi)^2}{2D^2} \tag{3.202}$$

For this case also considering a small value of x and Equation (3.202).

$$I = l.[\ln(\frac{\pi^2}{2D^2}\{x^2 + l^2\})] + 2x.\arctan\frac{1}{x} - 2l \tag{3.203}$$

3.4.15.3 Flow from uniform rainfall

The complex velocity for such condition is,

$$W_3 = u_3 + iv_3 = \frac{Rz}{D} \tag{3.204}$$

$u_3 = Rx/D$

where, w = u - iv, is uniform.

The horizontal component is uniformly distributed along any vertical. The complex potential is,

$$f_3(z) = \phi_3(x, y) + i\psi_3(x, y) = -\frac{Rz^2}{2D} + cons\tan t$$

3.205

On the upper boundary y=D

$$\phi_3(x, D) = \frac{Rx^2}{2D} + cons\tan t$$

3.206

$$\psi_3(x, D) = -Rx + cons\tan t$$

3.207

3.4.15.4 Formulation of drainage problem

First the steady rain seeping to an array of parallel equidistant drains or ditches will be investigated by linearization method (**Figure 3.28**). The complex potential which satisfies the boundary condition of **Figure 3.28** may be obtained from superposition of the elementary potentials previously developed.

For a drain as shown in **Figure 3.28a.**

$$f(z) = f_1(z) + f_3(z) = \frac{Q}{2\pi}\ln(\cosh\frac{\pi z}{D} - \cos\frac{\pi y_o}{D}) - \frac{Rz^2}{2D} + cons\tan t$$

3.208

and for a ditch,

$$f(z) = f_1(z) + f_3(z) = \frac{q}{2\pi}\int_0^l \ln(\cosh\frac{\pi z}{D} - \cos\frac{\pi\xi}{D})d\xi - \frac{Rz^2}{2D} + cons\tan t$$

3.209

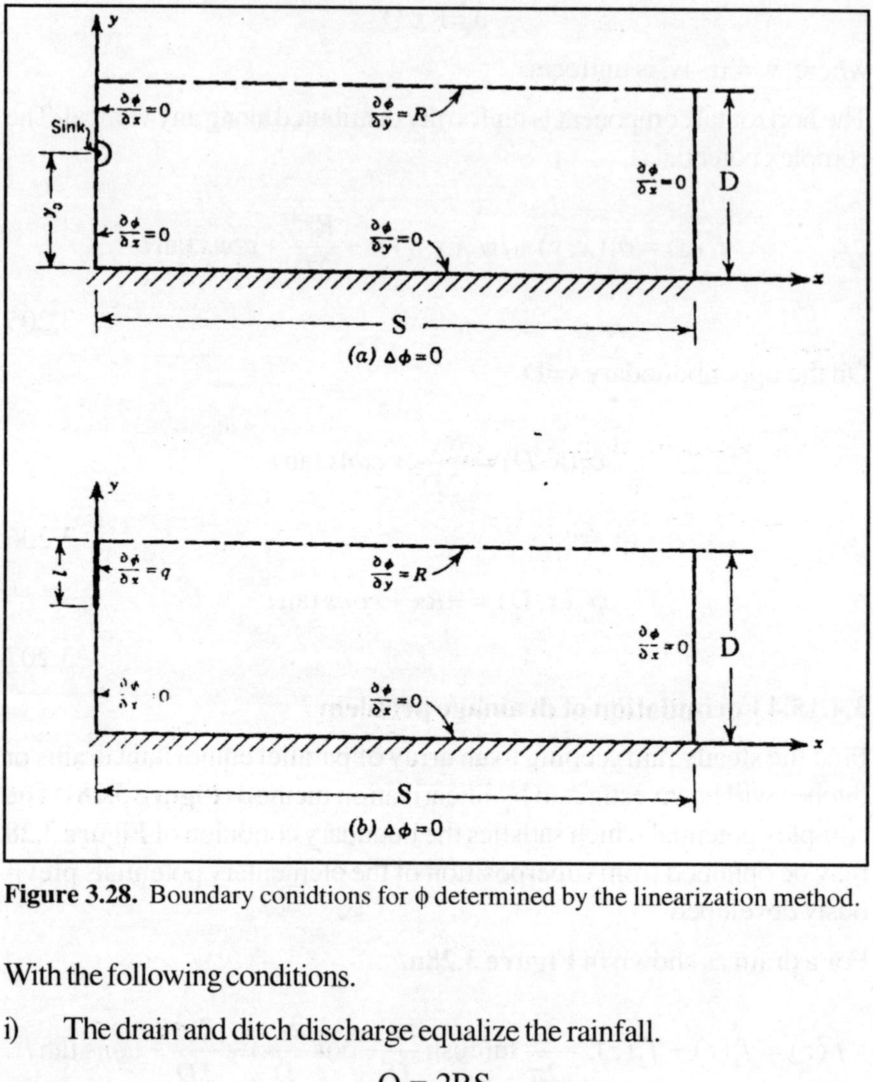

Figure 3.28. Boundary conidtions for ϕ determined by the linearization method.

With the following conditions.

i) The drain and ditch discharge equalize the rainfall.

$$Q = 2RS \tag{3.210}$$

$$q.\, l = 2RS \tag{3.211}$$

where,

S = semi drain spacing, L/2, [L]

ii) Pervious stratum is relatively thin, S>2D

This condition assures that at x=S, the velocity horizontal component will be zero and the boundary x=S, $0 \le y \le D$ will be a stream line.

iii) Drain radius and ditch width are small, r/D<<1

These conditions are fulfilled, the complete solution for the linearlized problem is possible and is easy to verify the complex potentials of Equation (3.208) and (3.209) satisfy all the boundary conditions of **Figure 3.28.**

Hence, free surface shape results accordingly to,

$$h_{(x)} = \frac{\phi_{(x,D)}}{K}$$

and from the value of the potential ϕ on the upper boundary y=D i.e. for a drain (From Equation (3.208), (3.209) and (3.198)).

$$h_{(x)} = \frac{\phi_{(x,D)}}{K} = \frac{\phi_{1(x,D)}}{K} + \frac{\phi_{3(x,D)}}{K} = \frac{RS}{K}\ln(\cosh\frac{\pi x}{D} + \cos\frac{\pi y_o}{D}) - \frac{Rx^2}{2KD} + constant \quad 3.212$$

For a ditch Equations (3.209), (3.207) and (3.200)

$$h_{(x)} = \frac{\phi_{(x,D)}}{K} = \frac{\phi_{2(x,D)}}{K} + \frac{\phi_{3(x,D)}}{K}$$

$$h(x) = \frac{RSD}{K\pi^2 l}\int_0^{\pi l/D} \ln(\cosh\frac{\pi x}{D} + \cos\mu)d\mu - \frac{Rx^2}{2KD} + constant \quad 3.213$$

where,

μ = integration variable

However, at a maximum distance x=2D, the horizontal velocity component will be uniformly distributed on vertical ($\partial u/\partial y \cong 0$) and potential of x only.

Consequently in the region of uniform flow defined by 2D<x<S, the shape of free surface will be better approximated by a solution based on the ellipse equation than by linearized method.

In the region 0<x<2D, the free surface position will be derived by linearized method and in the region 2D<x<S by Dupuit assumption (ellipse equation).

3.4.15.5 Homogeneous soil

(a) Drain located at water table

Figure 3.29 shows this case where yo=D. In the strip r<x<2D free surface will be determined by the linearized method. It is assumed that drain is half full and water table is located at y=D in the drain using Equation (3.212) and condition.

$y_o = D, \qquad x = r, \qquad \phi = KD$ 3.214

The equation of free surface will be

$$h_{(x)} - D = \frac{RS}{\pi K} \ln \frac{\cosh \frac{\pi x}{D} - 1}{\cosh \frac{\pi r}{D} - 1} - \frac{Rx^2}{2KD} \qquad 3.215$$

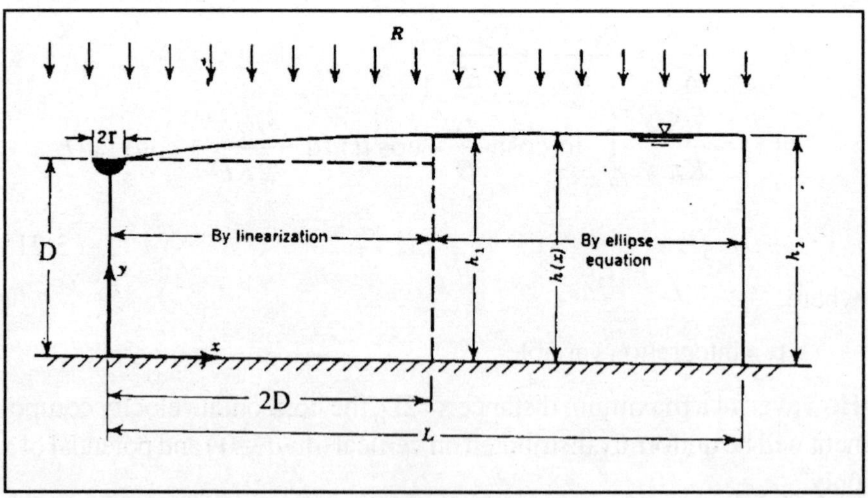

Figure 3.29. Drain location at a water table.

Drawdown at x=2D (neglect l against cosh πx/D) will be.

$$h_1 - D = \frac{2R}{K}(S - D) - \frac{RS}{\pi K}\ln 2(\cosh\frac{\pi r}{D} - 1)$$

3.216

In the strip 2D<x<S, ellipse equation can be used.

$$h_{(x)}^2 = h_2^2 - \frac{R}{K}(S - x)^2$$

3.217

Total drawdown between the drain (x=r, y=D) and the central point between the drain (x=S, y=h_2) can be obtained after combining Equation (3.216) and (3.217).

$$h_2^2 - D^2 = \frac{R}{K}[S(S - 2\alpha_1 D)]$$

3.218

$$\alpha_1 = \frac{1}{\pi}\ln 2(\cosh\frac{\pi r}{D} - 1)$$

3.219

The value of α_1 can be determined by using Nomograph as shown in **Figure 3.30**.

(b) Drain below a water table

When yo <D in the strip 0<x<2D, the same relationship (Equation 3.212) will furnish the solution, with the condition that drain water table is located at (**Figure 3.31**).

h = D, or $\phi_{(0,D)} = KD$ 3.220

The equation of water table in the region 6<x<2D will be

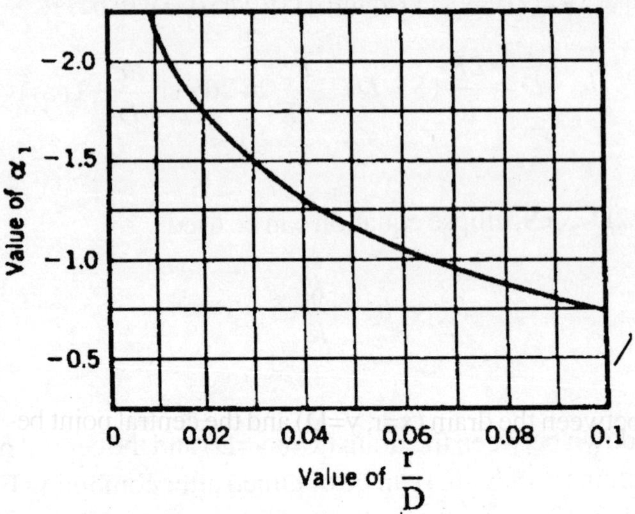

Figure 3.30. Nomographs for determining α_1

$$h_{(x)} - H = \frac{SL}{K} \ln \frac{\cosh \frac{\pi x}{D} + \cos \frac{\pi y_o}{D}}{1 + \cos \frac{\pi y_o}{D}} - \frac{Rx^2}{2KD}$$

3.221

The drawdown at x=2D (after neglecting $\cos \pi y_o/D$ relative to $\cosh \pi x/D$) will be

$$h_1 - H = \frac{2R}{K}(S - D) - \frac{SR}{\pi K} \ln 2(1 + \cos \frac{\pi y_o}{D})$$

3.222

Combining Equations (3.221) with ellipse equation

$$h_2^2 - D^2 = \frac{R}{K}[L(L - 2\alpha_2 D)]$$

3.223

$$\alpha_2 = \frac{1}{\pi} \ln 2(1 + \cos \frac{\pi y_o}{D})$$

3.224

The values of α_2 can be also obtained from **Figure 3.32.**

(c) Flow to a ditch

Figure 3.33 shows flow pattern schematically towards ditch.

Equation (3.213) with condition that x=δ, y = H and ϕ=KD, yield for the water table for 0<x<2D.

$$h_{(x)} - D = \frac{RDS}{\pi^2 lK} \int_0^{\frac{\pi l}{D}} \ln \frac{\cosh \frac{\pi x}{D} - \cos \mu}{\cosh \frac{\pi \delta}{D} - \cos \mu} d\mu - \frac{R(x^2 - \delta^2)}{2DK}$$

3.225

for large value of x (x>1.5D), neglecting δ relative to x and cos μ relative to cosh πx/D Equation (3.225) becomes.

$$h_{(x)} - D = \frac{Rx}{2K}(2S - x) - \frac{RS}{K}(\frac{\ln 2}{\pi} + \frac{D}{\pi^2 l} I)$$

3.226

where,

$$I = \int_0^{\frac{\pi l}{D}} \ln(\cosh \frac{\pi \delta}{D} - \cos \mu) d\mu$$

3.227

Water table drawdown at x=2D will be

$$h_1 - D = \frac{2R}{K}(S - D) - \frac{RS}{\pi K}(\ln 2 + \frac{D}{\pi\, l} I)$$

3.228

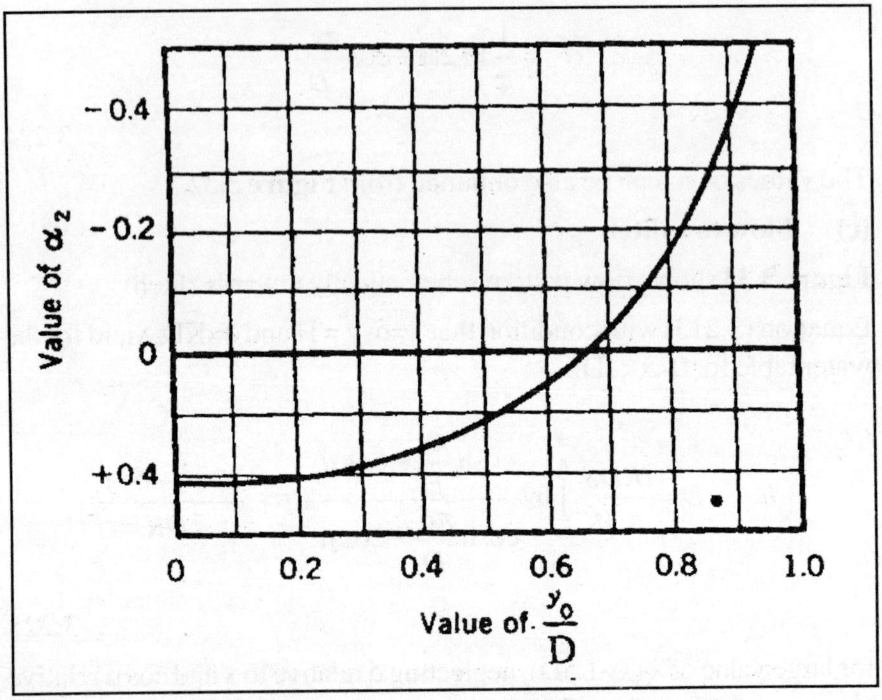

Figure 3.31. Drain located below water table.

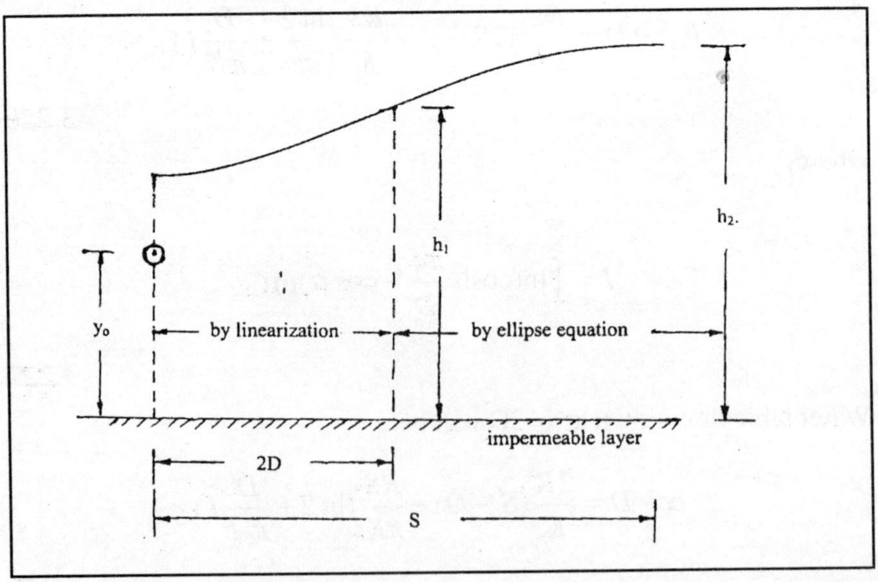

Figure 3.32. Nomograph for determining α_2

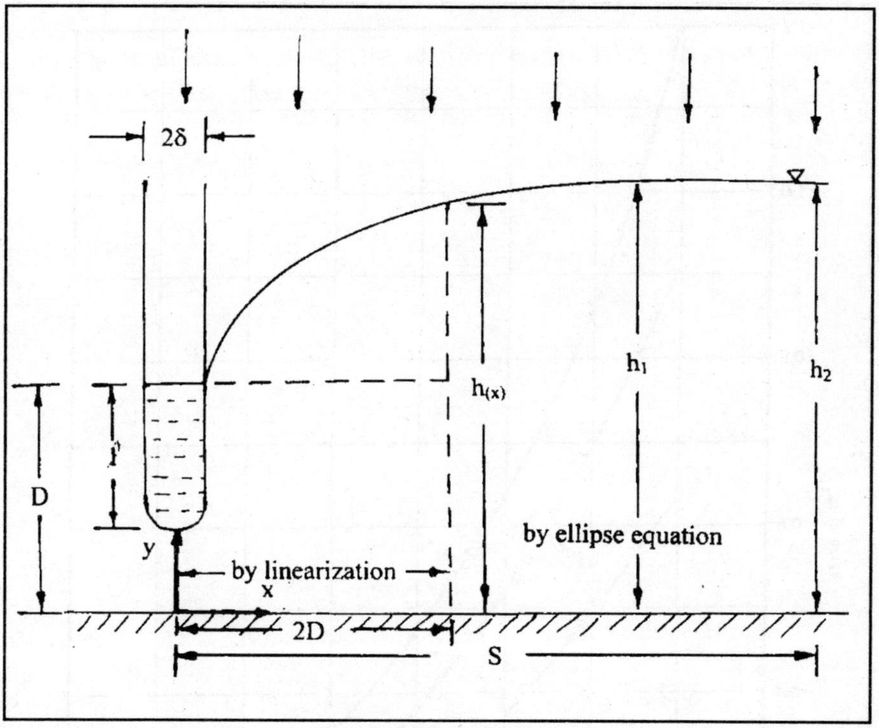

Figure 3.33. Seepage of rainfall towards ditch.

Combining Equation (3.228) with ellipse equation valid for 2D<x<S, the final result for the maximum drawdown will be.

$$h_2^2 - D^2 \cong \frac{R}{K}[S(S - 2\alpha_3 D)]$$

3.229

$$\alpha_3 = \frac{1}{\pi}(\ln 2 + \frac{D}{\pi l} I)$$

3.230

The value of α_3 can be determined with the help of nomograph as shown in **Figure 3.34**.

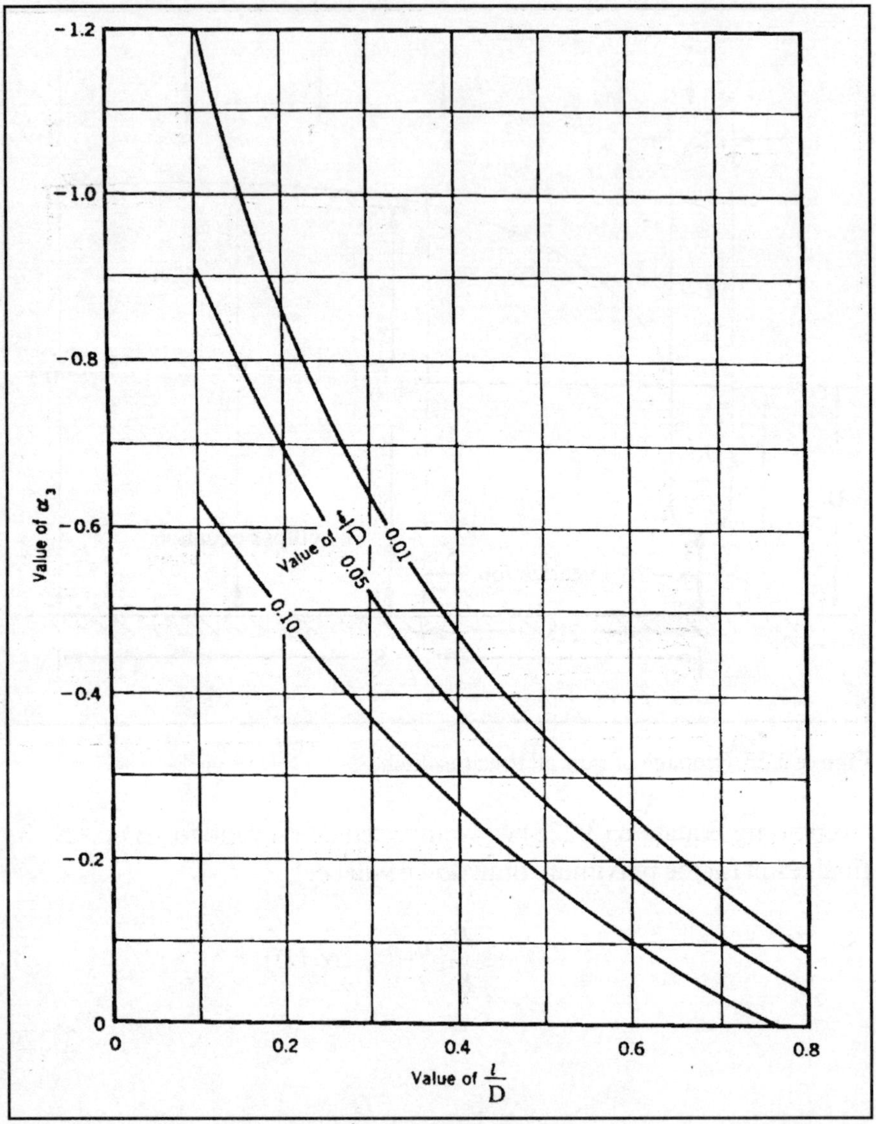

Figure 3.34. Nomograph for determination of α_3

3.4.16 Application of Dagan's drain spacing equation in non-isotropic soil

In many cases horizontal hydraulic conductivity K_x is not same as vertical hydraulic conductivity K_y. The ellipse equation cannot account this condition since it assumes horizontal flow only which is influenced by K_x. Solution of flow problem in an isotropic porous medium may be transferred to non-isotropic medium by use of an approximate scale. Namely it is possible to pass from an equivalent isotropic model to the non-isotropic prototype and reciprocally by the transformation.

$$\frac{y_e}{y_p} = 1 \tag{3.231}$$

$$\frac{Q_e}{Q_p} = 1 \tag{3.232}$$

$$\frac{\phi_e}{\phi_p} = 1 \tag{3.233}$$

$$\frac{x_e}{x_p} = \sqrt{\frac{K_{yp}}{K_{xp}}} \tag{3.234}$$

$$\frac{R_e}{R_p} = \sqrt{\frac{K_{yp}}{K_{xp}}} \tag{3.235}$$

$$K_e = \sqrt{K_{xp}.K_{yp}} \tag{3.236}$$

In which subscript e refers to the equivalent isotropic model, and subscript p to non-isotropic prototype. These relations involve geometrical stretching. Hence if the boundaries are only vertical or horizontal straight lines, the geometry of the non-isotropic prototype will be similar to the equivalent isotropic model. However a circular drain may be transferred into an elliptical one. Hence to obtain a circular drain in non-isotropic prototype, the problem of flow to an elliptical drain must be solved in the isotropic equivalent model.

i) Drain located at water table (y_o=D)

In general K_x>K_y, the scale changes of Equation (3.231 to 3.236) will transform the equivalent isotropic model into a shorter one and the semi-circular drain into an elliptical perimeter may be considered as an equipotential around a sink, similar to a ditch. Hence, maximum drawdown caused by a drain of radius r, in a non-isotropic layer of height D and length S with hydraulic conductivity K_x and K_y and rain intensity R will be equal to that caused by a ditch of width,

$$\delta_e = r\sqrt{\frac{K_y}{K_x}} \quad 3.237$$

$$l_e = r \quad 3.238$$

in an isotropic layer of height D and length S,

$$S_e = S\sqrt{\frac{K_y}{K_x}} \quad 3.239$$

hydraulic conductivity,

$$K_e = \sqrt{K_x K_y} \quad 3.240$$

and rain intensity,

$$R_e = \sqrt{K_x / K_y} \quad 3.241$$

Hence Equation (3.229) yield.

$$h^2 - D^2 = \frac{R}{K_x}[S(S - 2\sqrt{\frac{K_x}{K_y}}\alpha_3 D)]$$

3.242

in which all the symbols are related to the non-isotropic case and α_3 corresponds to the values.

$$\frac{\delta_e}{D} = \frac{r}{D}\sqrt{\frac{K_y}{K_x}}$$

3.243

$$\frac{l_e}{D} = \frac{r}{D}$$

3.244

Generally $\delta_e/D < 0.10$ and $l_e/D < 0.05$

Hence the integral I, which enters in the expression of α_3 may be approximated with the result of Equation (3.203) and Equation (3.229), the result is,

$$\alpha_3 = \frac{1}{\pi}[2\ln\frac{\pi r}{D} + \ln(1 + \frac{K_y}{K_x}) + 2\sqrt{\frac{K_y}{K_x}}\arctan\sqrt{\frac{K_x}{K_y}} - 2]$$

3.245

(ii) Application to ditches

Same technique may be used for the solution of the problem of rain seeping towards a ditch in a non-isotropic layer. Equation (3.229) can be changed to,

$$h^2 - D^2 = \frac{R}{K_x}[L(L - 2\sqrt{\frac{K_x}{K_y}}\alpha_3.D)]$$

3.246

α_3 can be determined for,

$$\frac{\delta_e}{D} = \sqrt{\frac{K_y}{K_x}}..\frac{\delta}{D}$$

3.247

$$\frac{l_e}{D} = \frac{l}{D}$$

3.248

Problem 3.8

After a detailed study of a waterlogged area in Tarai belt of Nainital, following information were collected in order to design a subsurface drainage system in the area.

K=2 m/day

R = 10 mm/day

D = 2 m

r = 0.10 m

h_2= h+D=2.3 m

Calculate drain spacing, L to overcome the steady state recharge.

Solution 3.8

1. Following drain spacing formula will be applicable for the given situation.

$$h^2{}_2 - D^2 = \frac{R}{K}[S(S - 2\alpha_1 D)]$$

$$\alpha_1 = \frac{1}{\pi}\ln 2(\cosh\frac{\pi r}{D} - 1)$$

2. Calculate value of α_1 as,

α_1 = 1/π ln 2 (cosh πx0.10/2 –1) = -1.178

3. Calculate drain spacing as below,

$$2.3^2 - 2^2 = 0.01/2\ [S(S+2x1.178x2)]$$

$$258 = S^2 + 4.712\ S$$

Solving the above quadratic equation for S, which comes out to be,

$$S= 13.88$$

Therefore, L =2S≅27.8 m

Problem 3.9

A drainage detailed investigation of a waterlogged area in Sharda Sahayak Canal Command in Raebareli district of UP were made and following information were collected in order to design a subsurface drainage system in the area.

K=1.5 m/day

D = 2.5 m

r = 0.12 m

Water level midway between the drains to be maintained at 0.5 m above drainage level. Water balance study indicated that seepage has to be removed at rate of 0.01 m per day. What should be the spacing between the drain?

Solution 3.9

1. Calculate h_2= h+D=0.5 + 2.5 m =3.0 m
2. Following drain spacing formula will be applicable for the given situation.

$$h^2{}_2 - D^2 = \frac{R}{K}[S(S - 2\alpha_1 D)]$$

$$\alpha_1 = \frac{1}{\pi}\ln 2(\cosh\frac{\pi r}{D} - 1)$$

2. Calculate value of α_1 as,

$$\alpha_1 = 1/\pi \ln 2\ [\cosh\ (\pi x0.12/2) - 1] = -1.204$$

3. Calculate drain spacing as below,

$$3.0^2 - 2.5^2 = 0.01/2\ [S(S+2x1.204x2.5)]$$

$$412.5 = S^2 + 6.02\ S$$

Solving the above quadratic equation [coefficient of S^2, a=1, coefficient of S, b=6.02 and constant, C = 412.5] for S.

$$S = \frac{-b+\sqrt{b^2-4aC}}{2a} = \frac{-6.02+\sqrt{(-6.02)^2-4\{1.(-412.5)\}}}{2} = 17.52m$$

$$S = 17.52$$

Therefore, L =2S≅35 m

Problem 3.10

In a non-isotropic soil when the drain located at water table (yo=D), following information are available.

K_x = 2 m/day

K_y = 0.4 m/day

R = 10 mm/day

D = 2 m

r= 0.10 m

h_2 = 2.3 m

What will be the spacing between the drains.

Solution 3.10

1. Transform the problem into isotropic equivalent model as follows:

   $\delta_e/D = r/D\sqrt{(K_y/K_x)} = 0.1/2\ \sqrt{(0.4/2)} = 0.0224$

2. Calculate the term l_e/D = r/D =0.1/2 =0.05.

3. Make use of Equation (3.245) and calculate the value of α_3 as -1.43.

4. Make use of Equation (3.242) and calculate half spacing S as under.

$2.3^2 - 2^2 = 0.01/2\ [S(S + 2x\sqrt{(2/0.4)} \times 1.43 \times 2]$

Solving the above quadratic equation for S one will get the value of S=10.9 m.

5. Calculate the full drain spacing as L=2S= 21.8 m.

Problem 3.11

Estimate the spacing between the ditches with the following information.

$K_x = 2$ m/day

$K_y = 0.4$ m/day

R = 10 mm/day

D = 2 m

l= 1 m

$h_2 = 2.3$ m

Solution 3.11

1. Calculate the term $\delta_e/D = \delta/D\ \sqrt{(K_y/K_x)} = 0.2/2\ \sqrt{(0.4/2)} = 0.0448$.
2. Calculate l/D = ½ = 0.5.
3. Estimate the value of α_3 from **Figure 3.34,** which works out to be 0.32.
4. Make use of Equation (3.246) and calculate half spacing S as under,

   $2.3^2 - 2^2 = 0.01/2\ [S(S + 2x2.4\ x0.32 \times 2]$

   Solving above quadratic equation for S, the value may be obtained as 14.65 m.
5. Spacing between ditches will be thus, L=2S= 29.3 m.

3.5 Unsteady State Drainage Theories

In areas with periodic irrigation or high intensity rainfall, the assumption of a steady recharge is no longer justified. Under these conditions non-steady state solution of the flow problem must be applied. Many designers feel that an equation that describes the movement of the water table through the soil will provide a more accurate method of design. The movement of water table through the soil is a transient condition. The hydraulic head at any point in the soil is not constant; it is changing with time. Therefore, the situation is called the "transient" state as opposed to the "steady state" in which hydraulic head does not vary with time.

In humid areas like Assam and Tarai belt of Northern India all along the foot hill of the Himalayan mountainous region, precipitation pattern are relatively steady in nature, that is, prolonged, medium intensity rain spells are frequently encountered. Under such conditions steady state solutions can be successfully applied.

Steady state drainage criteria, however, are not available or are not satisfactory in all areas, in fact, steady state conditions are the exception rather than the rule. Many designers of the drainage system have felt that the formulas which describe the water table in equilibrium with rainfall or irrigation water do not conform to the field situation where the water table is constantly moving through the soil due to periodic or high intensity rainfall. The rise and fall of the water table in irrigated areas generally follows a cycle which is related to the application of irrigation water during the growing season and the termination of irrigation water use in the off season. An equation that describes the movement of water table through the soil would provide a more accurate method of design. Such a movement of the water table through the soil is called a 'non-steady' state or 'transient' condition.

Non-steady/time varying or transient flow problems are more tedious to solve than time invariant problems. Non-steady state solutions however, are indispensable when actual, non steady water table elevations and drain discharges, as obtained from field data must be evaluated.

In following section we will review some of the drainage equations derived for non-steady conditions and indicate their usefulness in application of field drainage design.

An equation that has widely been used for transient flow problems is based on Dupuit-Forchheimer assumptions as discussed earlier.

3.5.1 Governing equation

As outlined earlier, for the D-F assumptions to be reasonable, their application should be restricted to flow system that is shallow relative to its aerial extent and that exhibit a relatively small curvature of the ground water table. The equation referred to above describes an approximation-horizontal flow in a vertical prism in a flow region, bounded above by the free ground surface and below by an impermeable layer as shown in **Figure 3.35**. **Figure 3.35** approximate horizontal flow in a space element of fluid as an assumption for deriving Forchheimer's linearized continuity equation.

The equation can be derived by combining Darcy's law with the water balance for such a prism, located between parallel drains. In this D-F conditions flow is considered in one dimension or direction only; the x-direction, of **Figure 3.36**. The changes of flow in the x-direction must equal the change in the quantity of water stored in the column considered. This change in storage is reflected either in a rise or a drop of the ground water table. The change of storage is defined as,

$$\Delta s = f\,\Delta h \qquad 3.249$$

where,

Δs = change in water stored per unit surface area during time step, [L]

f = the effective porosity of soil, dimensionless [L^3/L^3]

Δh = change in ground water table level during time step, [L]

The effective porosity also referred to as 'drainable porosity', represents the total fraction of the soil volume which is drained as the ground water table goes down. The use of a constant value for, however, is a simplification of actual conditions because the volume of water which is drained from the soil increases gradually with an increase of suction or tension. The effective porosity is far from constant, depending at least on the proximity of the water table to the soil surface and on the rate and direction of the water table movement. The result of this simplification is difficult to assess but accuracy of the approximate solutions is found to be good enough for practical purposes.

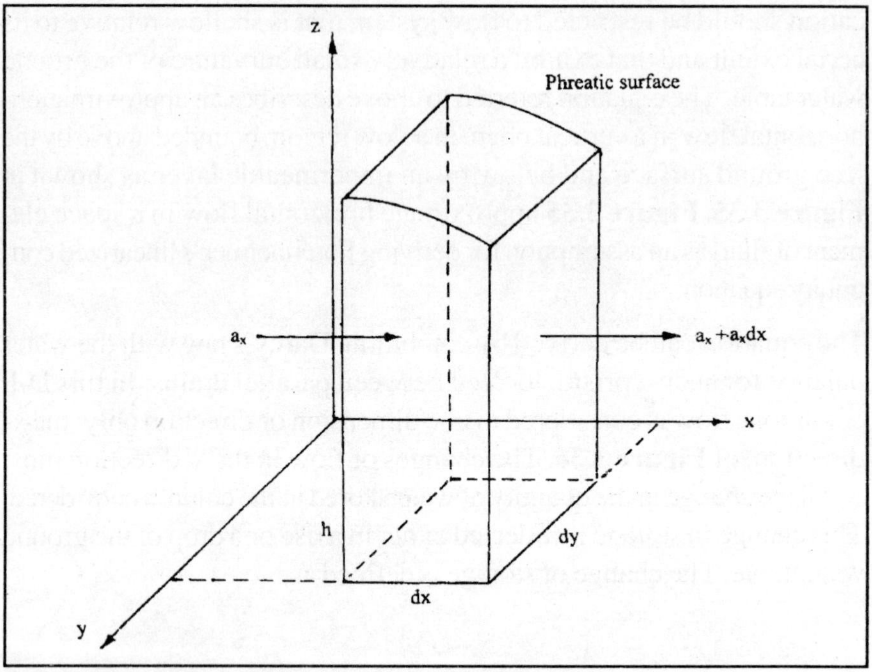

Figure 3.35. Approximate horizontal flow in a space element of fluid as an assumption for deriving Forchheimer's linearized continuity equation.

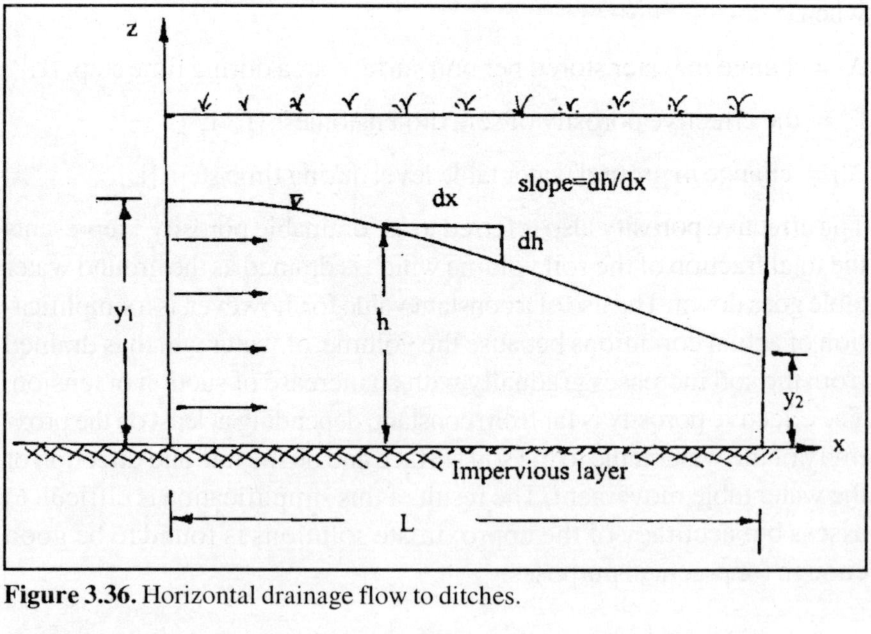

Figure 3.36. Horizontal drainage flow to ditches.

Combining Darcy's law with continuity equation yields, after rearrangement.

$$\frac{\partial h}{\partial t} = \frac{KD}{f} \cdot \frac{\partial^2 h}{\partial x^2} + \frac{R}{f}$$

3.250

where,

f = drainable porosity, [L^3/L^3]

t = time, [T]

Other notations are as earlier.

Equation (3.250) is called the Boussinesq equation, which describes the position of water table under non-steady state conditions. When recharge rate R equal zero the equation reduces to,

$$\frac{\partial h}{\partial t} = \frac{KD}{f} \cdot \frac{\partial^2 h}{\partial x^2}$$

3.251

This is the equation that must be solved for non-steady state condition of the moving water table. This equation is often referred to as the standard heat flow equation, because it is identical in form to the differential equation that applies to many heat flow problems. The methods for solving the heat flow equation under different initial and boundary conditions are extensively discussed in **Carlslaw and Jaeger (1959)** and the methods used by them can be used as well for our problems.

3.5.2 Glover-Dumm drain spacing equation

Dumm (1954) used a solution for Equation (3.251) found by Glover who assumed an initial horizontal ground water table at a certain height above drain level. The solution describes the lowering of ground water table, which does not remain horizontal, as a function of time. **Figure 3.37** shows the geometry of the Glover drain equation.

3.5.2.1 Derivation using continuity equation

The initial horizontal ground water table is thought to have been the result of an instantaneous rise caused by rainfall or irrigation, which instantly recharge the ground water.

Following assumptions were made for derivation of **Glover (1954)** equation :

1. Normal irrigation at regular interval.
2. Uniform or stratified soils overlying a slowly permeable stratum or barrier layer.
3. Permeability of the profile equal to the weighted average permeability of the soils above the barrier layer.
4. The drainage requirement, or excess water to be disposed of, is the

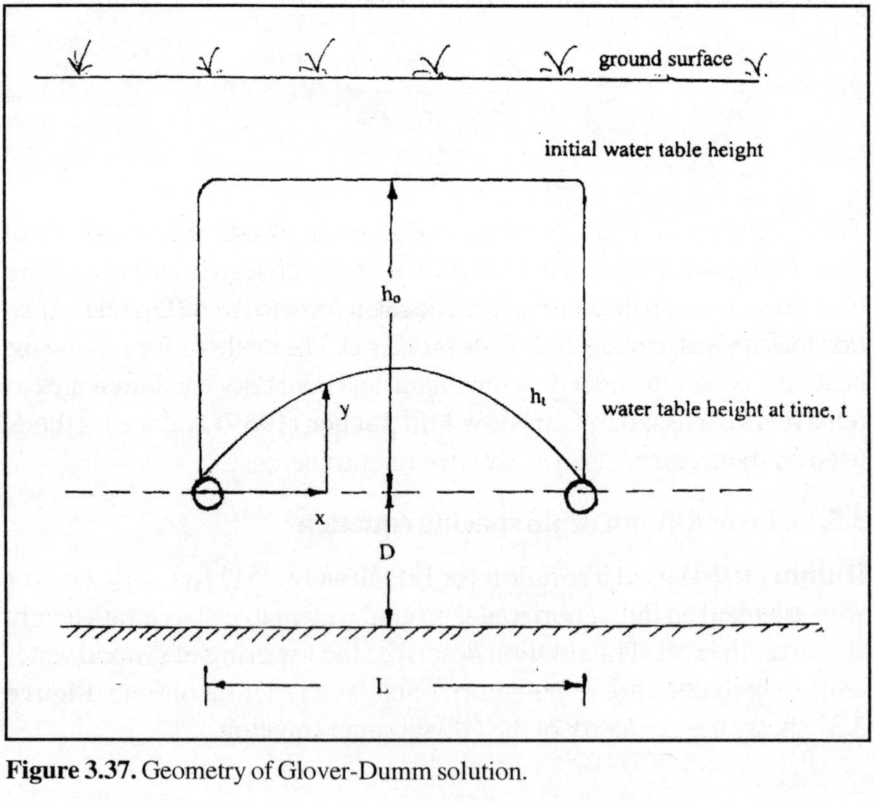

Figure 3.37. Geometry of Glover-Dumm solution.

difference between the net input of water and available moisture storage in that portion of the profile which must be kept clear of gravity or non-capillary water in order to provide for root growth and salinity control.

5. Excess water must be removed between irrigations in order to maintain the water table below root zone. This would require that sufficient storage space for gravity water be provided below the root zone for the inflow or excess water from the following irrigation.
6. Water moving through the soils is in a transitory state rather than a steady state, i.e. the head available to push the water towards the drains varies with time.
7. D-F assumptions are valid.

Glover used the differential Equation (3.251) which may be rewritten as below for obtaining a non-steady solution.

$$\frac{\partial h}{\partial t} = \frac{KD_a}{f} \cdot \frac{\partial^2 h}{\partial x^2} \qquad 3.252$$

where,

D_a = $(D+1/2\ h_o)$ which represents an average thickness of aquifer available for carrying water to the drains, [L]

Following boundary conditions were used for obtaining solution :

$$h = 0 \text{ at } x=0$$

$$h = 0 \text{ at } x=L \qquad 3.253$$

$$h = ho, \text{ when } t=0 \text{ and } 0<x<L$$

General solution of Equation (3.252) can be written as,

$$h = u + w \qquad 3.254$$

where,

u is for steady state and w is for unsteady state.

For getting the solution of steady state condition substitute the Equation (3.254) in Equation (3.252) one gets,

$$\frac{\partial^2 u}{\partial x^2} = 0 \tag{3.255}$$

Boundary condition for steady state conditions can be transformed as:

$$u(0) = 0$$
$$u(L) = 0 \tag{3.256}$$

Now upon integration of Equation (3.255) one get,

$$\frac{\partial u}{\partial x} = C_1 \tag{3.257}$$

$$u = C_1 x + C_2 \tag{3.258}$$

Substitute the boundary condition (3.256) in Equation (3.258) one will get C_1=0, C_2=0 and u=0.

Therefore solution of Equation (3.252) becomes,

$$h = w \tag{3.259}$$

and general solution of transient condition can be obtained as follow **(Carlslaw and Jaeger, 1959)**.

$$h = w = \sum_{n=1}^{\infty} A_n e^{-(\frac{n^2\pi^2 KD}{fL^2})t} .\sin\frac{n\pi x}{L} dx \tag{3.260}$$

where,

$$A_n = \frac{2}{L}\int_0^L f_{(x)} \sin\frac{n\pi x}{L} dx$$

3.261

Since initially water table was assumed to be constant therefore $f(x)=h_o$. Now A_n can be worked out as follows :

$$A_n = -\frac{2}{L}\int_0^L h_o \sin\frac{n\pi x}{L} dx$$

$$A_n = -\frac{2h_o}{L}\int_0^L \sin\frac{n\pi x}{L} dx$$

$$= -\frac{2h_o}{L}[-\frac{L}{n\pi}\cos\frac{n\pi x}{L}]_0^L$$

$$= +\frac{2h_o}{L}[\frac{L}{n\pi}\cos(0) - \frac{L}{n\pi}\cos\frac{n\pi L}{L}]$$

$$= -\frac{2h_o}{L}[\cos(n\pi) - \cos(0)]\frac{L}{n\pi}$$

$$= -\frac{2h_o}{n\pi}[(-1)^n - 1] = -\frac{2h_o}{n\pi}[-2]$$

3.262

$$A_n = \frac{4h_o}{n\pi}$$

3.263

Equation (3.260) can be written as,

$$h_{(x,t)} = \sum_{n=1}^{\infty}\frac{4}{n}\frac{h_o}{\pi}e^{-(\frac{n^2\pi^2 KD_a}{fL^2})t} . \sin\frac{n\pi x}{L}$$

3.264

$$h_{(x,t)} = \frac{4h_o}{\pi} \sum_{n=1}^{\infty} \frac{1}{n} e^{-n^2 \alpha t} . \sin \frac{n \pi x}{L}$$

3.265

where,

$$\alpha = \frac{\pi^2 K D_a}{f L^2}$$

3.266

The term α is also called as reaction factor. The larger is the α, faster will be the drainage system response.

Equation (3.265) gives the shape of water table at any time t and at location x. Because we are mainly interested in the midway water table height between the drains, we substitute x=1/2 L into Equation (3.265).

$$h_t = \frac{4h_o}{\pi} \sum_{n=1}^{\infty} \frac{1}{n} e^{-n^2 \alpha t} . \sin \frac{n \pi}{2}$$

3.267

For even numbers of 'n' sin term will be equal to zero. Only odd number will have non-zero value. For getting a positive solution n should be selected as, n=1,-3,5,-7,.....

Therefore Equation (3.267) reduces to,

$$h_t = \frac{4h_o}{\pi} \sum_{n=1,-3,5,..}^{\infty} \frac{1}{n} e^{-n^2 \alpha t}$$

3.268

Obviously, the value of each term of Equation (3.268) decreases with increasing values of n. If t>0.2, the second next term will be comparatively small and can be neglected. Equation (3.268) then reduces to,

$$h_t = \frac{4h_o}{\pi} e^{-\alpha t} = \frac{4}{\pi} h_o e^{-\frac{\pi^2 K D_a}{f L^2} t}$$

3.269

or

$$\frac{4h_o}{\pi h_t} = e^{(\frac{\pi^2 K D_a}{f L^2}) t}$$

3.270

Taking $\log_e$ on both sides, one will arrive at,

$$\ln \frac{4h_o}{\pi h_t} = \frac{\pi^2 K D_a}{f L^2} t$$

$$L = \sqrt{\frac{\pi^2 K D_a t}{f . \log_e \frac{4}{\pi} \frac{h_o}{h_t}}}$$

when $D >> h_o$

3.271

The above equation is known as **Glover equation**.

Dumm (1960) assumed that the initial water table have the shape of fourth degree parabola. At time t=0, the water table has a shape given by the equation,

$$h = \frac{8h_o}{L^4}(L^3 x - 3L^2 x^2 + 4Lx^3 - 2x^4)$$

3.272

Dumm (1960) solved the differential Equation (3.252) under boundary conditions (3.253) assuming the initial shape of water table as fourth degree parabola. Solution obtained is as given below.

$$h_{(x,t)} = \frac{192h_o}{\pi^5}\sum_{0}^{\infty}\frac{n^2\pi^2-8}{n^5}e^{-(\frac{n^2\pi^2KD_a}{fL^2})t}.\sin\frac{n\pi x}{L}$$

3.273

At x=1/2 L, Equation (3.273) reduces to,

$$h_t = \frac{192h_o}{\pi^3}\sum_{n=1,3,5}^{\infty}(-1)^{\frac{(n-1)}{2}}.\frac{n^2-\frac{8}{\pi^2}}{\pi^5}.e^{-(\frac{n^2\pi^2KD_a}{fL^2})t}$$

3.274

Above equation can be also expressed in term of drain spacing considering only first term of the series as follow:

$$L = \sqrt{\frac{\pi^2KD_a t}{f.\log_e 1.16\frac{h_o}{h_t}}}$$

when $D>>h_o$ 3.275

This equation is known as Glover-Dumm equation, which is same as Glover equation except the change in shape factor from a value of $4/\pi$=1.27 to 1.16.

Because the Glover and Glover-Dumm equations do not take the radial flow into account, the average thickness of aquifer D_a is often replaced by the equivalent depth of Hooghoudt to account for the flow convergence in the vicinity of the drains. This substitution, which make Glover-Dumm applicable to pipe drainage, is justified because the flow paths for steady and transient flow may be considered at least similar though not identical. As a result Equation (3.266) becomes,

$$\alpha = \frac{\pi^2 Kd}{fL^2}$$

3.276

And Equation (3.271) and (3.275) change into,

$$L = \sqrt{\frac{\pi^2 K d_a t}{f.\log_e \frac{4}{\pi}\frac{h_o}{h_t}}}$$

when d>>h_o 3.277

$$L = \sqrt{\frac{\pi^2 K d_a t}{f.\log_e 1.16\frac{h_o}{h_t}}}$$

when d>>h_o 3.278

where,

d_a = d+ (1 /2) h_o

Although the above formula is derived under the assumption that h_o is small compared to d, we shall later see that if we use d_a = (d+ 1 /2 h_o) as the effective thickness of aquifer, they do give a fairly satisfactory approximation.

Glover (1954) also derived an equation for drain spacing when drains are resting on impervious layer. **Figure 3.38** shows the geometry of the flow condition for this case. If z represents the saturated thickness of aquifer at the distance x, measured horizontally form one line of drains, as shown in **Figure 3.38**, then the condition of continuity is,

$$\frac{\partial}{\partial x}(Kz\frac{\partial z}{\partial x}) = v\frac{\partial z}{\partial t} \quad 3.279$$

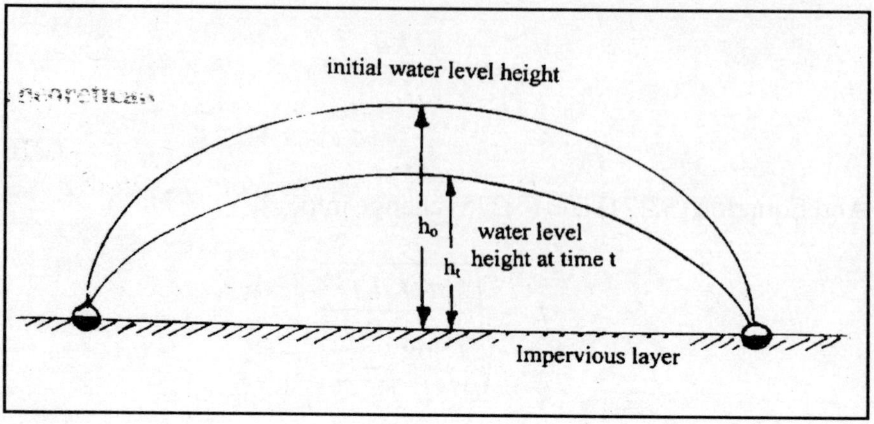

Figure 3.38. Geometry of drainage system when drains are resting at impervious layer.

A solution of this equation is written below,

$$\frac{z}{h_o} = wY \tag{3.280}$$

Where w is determined from relation,

$$\int_0^w \frac{wdw}{\sqrt{1-w^3}} \sqrt{3}\frac{x}{L} \tag{3.281}$$

And Y is given by,

$$Y = \frac{1}{\frac{9}{2}(\frac{Kh_o}{f}\frac{t}{L^2})+1} \tag{3.282}$$

If h_t represents the value of z at x=1/ 2 L at the time t, then since $Y=h_t/h_o$, a rearrangement of expression for Y yields,

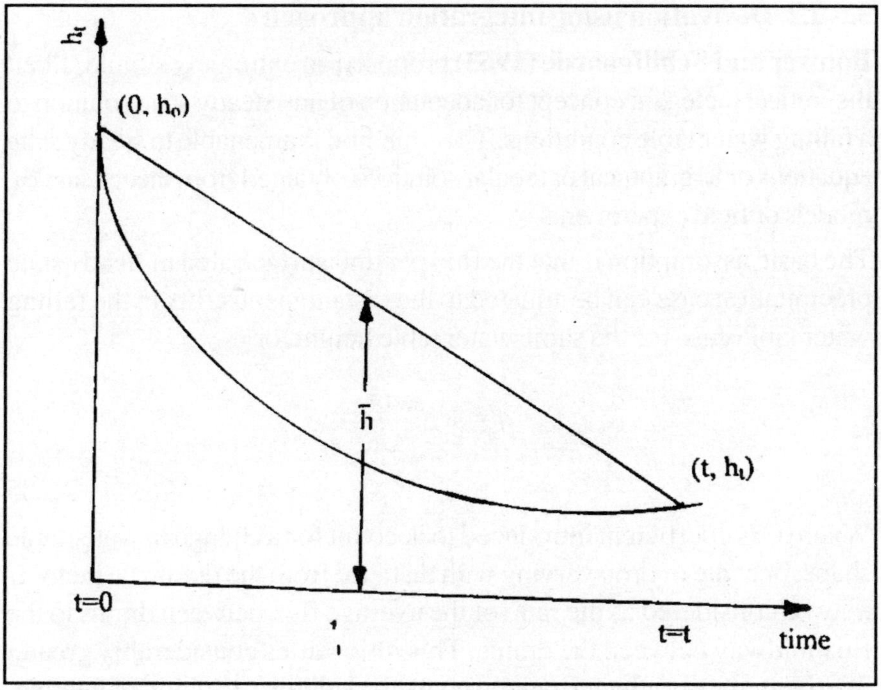

Figure 3.39. Water table draw down curve according to the Glover-Dumm equation.

$$t = \frac{2L^2}{9.\frac{Kh_o}{f}}\left(\frac{h_o}{h_t} - 1\right) \qquad 3.283$$

If this equation is solved for L, we get a formula for the drain spacing which is,

$$L = \sqrt{\frac{9Kh_o t}{2f\left(\frac{h_o}{h_t} - 1\right)}} \qquad 3.284$$

Equation (3.277), (3.278) and (3.284) can be derived using **Bouwer and Schilfgaarde (1963) approach.**

3.5.2.2. Derivation using integration approach

Bouwer and Schilfgaarde (1963) proposed an extremely simple, albeit theoretical inelegant concept for adaptation of any steady state solution to a falling water table conditions. Their method is amenable to steady state equations or to graphical or tabular solutions obtained from theory, analog models or field experiments.

The basic assumption is that the flux per unit surface area in steady state precipitation case can be equated to the instantaneous flux in the falling water table case for the same water table height, or

$$q = -fC\frac{dh}{dt} \tag{3.285}$$

Where C is coefficient introduced to account for a change in water table shape, or a rate of drop varying with distance from the drain, the factor C may be considered as the ratio of the average flux between drains to the flux midway between the drains. This ratio varies considerably greater than unity shortly after an initially ponded conditions; to unity for intermediate condition, to about 0.8 for small values of h (0.02 < h/L < 0.80). Unless circumstances warrant a more specific choice, experience has shown that a value of C=0.9 will yield good results.

Consider **Hooghoudt (1940)** steady state drain spacing formula reproduced as below.

$$q = \frac{8Kdh + 4Kh^2}{L^2}$$

or

$$q = \frac{8Kh}{L^2}(d + \frac{1}{2}h) \tag{3.286}$$

or,

$$q = \frac{8Kd_a h}{L^2} \tag{3.287}$$

where,

$d_a = (d+1/2\, h_o)$

Combining Equations (3.285) and (3.287) one will get,

$$-fC\frac{dh}{dt} = \frac{8Kd_a h}{L^2}$$

3.288

or

$$\frac{dh}{h} = -\frac{8Kd_a}{CfL^2}dt$$

3.289

Above equation can be integrated under the limit of t = 0, h = h_o and t = t, h = h_t.

$$\int_{h_o}^{h_t}\frac{dh}{h} = -\frac{8Kd_a}{CfL^2}\int_0^t dt$$

3.290

$$-\log_e h\Big]_{h_o}^{h_t} = +\frac{8Kd_a}{CfL^2}t\Big|_o^t$$

$$\log_e\frac{h_o}{h_t} = \frac{8Kd_a t}{CfL^2}$$

$$L = \sqrt{\frac{8Kd_a t}{Cf\log_e\frac{h_o}{h_t}}}$$

3.291

For C = 0.8, 8/.8=10≈π^2

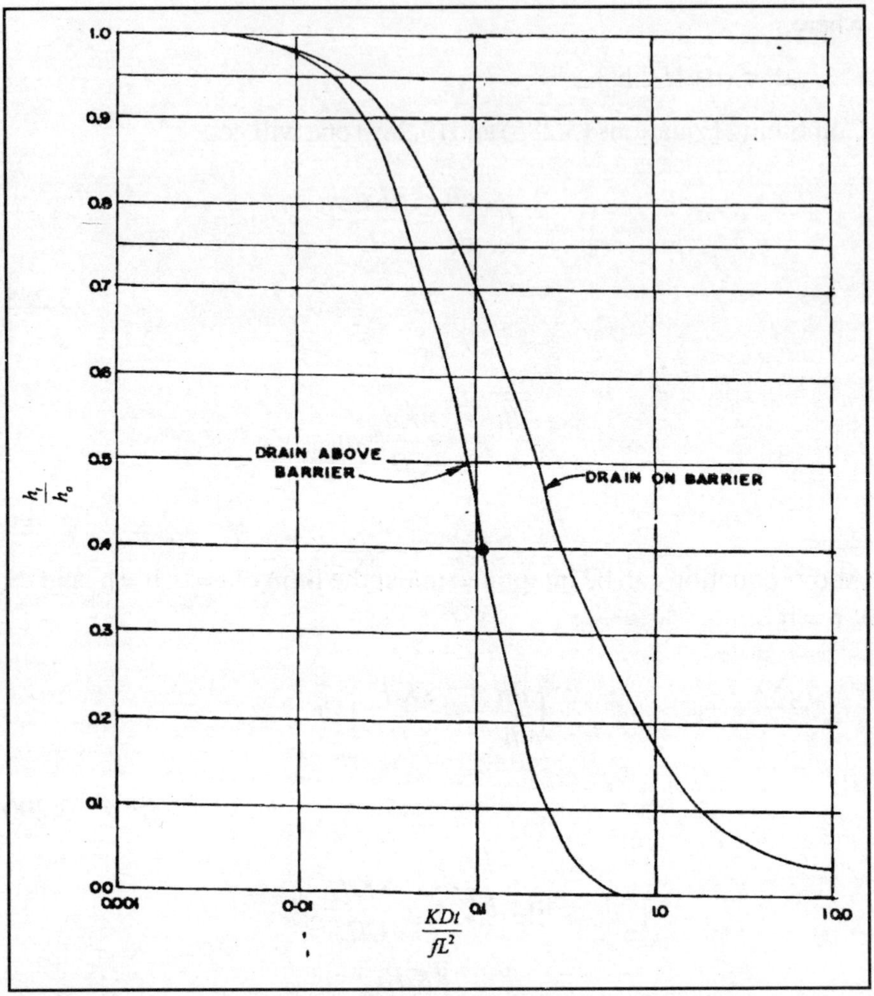

Figure 3.40. Graphical solution of the modified Glover-Dumm equation.

$$L = \sqrt{\frac{\pi^2 K d_a t}{f \log_e \frac{h_o}{h_t}}} \qquad 3.292$$

Which is similar to the Glover or Glover-Dumm equations only difference is of shape factor. Shape factor in Equation (3.292) is unity and initial shape of water table is elliptic whereas it was flat for the case of Glover equation and fourth degree parabola in case of Glover-Dumm equation. When flow above drain level is neglected Hooghoudt drain spacing formula reduces to,

$$q = \frac{4Kh^2}{L^2}$$

Combining above equation with Equation (3.285) will yield,

$$-fC\frac{dh}{dt} = \frac{4Kh^2}{L^2}$$

3.293

$$-\int_{h_o}^{h_t} \frac{dh}{h^2} = \frac{4K}{fCL^2}\int_0^t dt$$

3.294

Integrating Equation (3.294) under the limit of t =0, h=h_o and t=t, h=h_t, one will get.

$$\left(\frac{1}{h_t} - \frac{1}{h_o}\right) = \frac{4Kt}{CfL^2}$$

$$\left(\frac{h_o - h_t}{h_o h_t}\right) = \frac{4Kt}{CfL^2}$$

$$L^2 = \left[\frac{4Kt}{Cf\left(\frac{h_o - h_t}{h_o h_t}\right)}\right]$$

3.295

If C=0.9, the above equation reduces to,

$$L = \sqrt{\left[\frac{4.44Kh_o t}{f(\frac{h_o}{h_t} - 1)}\right]} \quad 3.296$$

Which is the same as Equation (3.284).

Instead of a water table elevation/ discharge criterion, as used for steady state equations, the Glover-Dumm equation requires a criterion as pre-determined water table drawdown in a certain period of time (h_o/h_t). In irrigated area drainage criteria states, what rate the water table midway between the drains has to drop after irrigation to a required new depth before the next irrigation starts.

3.5.3 Application of Glover-Dumm drain spacing equation

Example 3.12

In a canal irrigated area, irrigation water is applied at every 10 days. The field application losses, which percolate to the ground water are 25 mm after each irrigation, and are recharging as an instantaneous recharge, R_1=0.025 m. Effective drainable porosity 'f' of the area is 0.05, which causes instantaneous rise of water table, $\Delta h = R_1/f = 0.025/0.05 = 0.5$ m.

The maximum permissible height of the water table is set at 1 m below the ground surface. The drain level is chosen at 1.8 m below the soil surface. The water table must be lowered by h=0.5 m in the next 10 days, or else, with the subsequent irrigation to safe guard the crop, it will rise to above 1.0 m below the soil surface. If the depth to an impermeable layer is 9.5 m below soil surface and saturated hydraulic conductivity K= 1 m/ day, calculate the spacing between the drains for a drain pipe radius of 0.10 m.

Solution 3.12

Initial water table height above drain axis (hydraulic head) midway between the drains = 1.8 m – 1.0 m = 0.8 m

Water table height above drain axis after 10 days, h_{10} = 0.8 m – 0.5 m = 0.3 m

$K = 1.0$ m/day

Depth of impervious layer below drain axis, D = 9.5 m – 1.8 m = 7.7 m

$t = 10$ days

$r_o = 0.10$ m

Substituting above data into the Equation (3.278) will give,

$$L = \pi \left[\frac{1.0xdx10}{0.05} \right]^{\frac{1}{2}} \left[\ln \frac{1.16x0.8}{0.3} \right]^{-\frac{1}{2}}$$

$$L = 41.8\ \sqrt{d}\ \text{m}$$

Trial I: Assume value of drain spacing L = 80 m and read from **Figure 3.16** of van Beers nomograph for equivalent depth against the value of L= 80 m and D/u = $D/\pi r_o$ = 7.7/πx0.10 =25, which comes out to be, d= 4.4 m. Substitute this value of equivalent depth, d, spacing can be calculated as,

L = 41.8 = 41.8 √4.4 = 88 m. The calculated value of L is higher than assumed value. One has to increase the assumed value of L in second trial.

Trial II: Assume drain spacing L = 100 m.

Read the value of equivalent depth to impervious layer below drain axis from the **Figure 3.16** for D/u=25 and assumed drain spacing L =100 m, which came out to be, d = 4.8 m.

Substitute this value of equivalent depth in L = 41.8 √d = L = 41.8 √4.8 = 92 m. The calculated value of drain spacing is less than the assumed value hence the third trial assumed value of drain spacing should be reduced.

Trial III: Assume drain spacing L = 92 m.

Read the value of equivalent depth to impervious layer below drain axis from the **Figure 3.16** for D/u=25 and assumed drain spacing L = 92 m, which came out to be, d = 4.7 m.

Substitute this value of equivalent depth in L = 41.8 √d = L = 41.8 √4.7 = 90 m. The calculated value of drain spacing is almost equal to the assumed value of drain spacing, hence should be adopted finally.

Above problem may be also solved alternatively as shown below.

Compute L_o as follow,

$$L_o = \pi \left[\frac{KDt}{f}\right]^{\frac{1}{2}} \left[\ln 1.16 \frac{h_o}{h_t}\right]^{-\frac{1}{2}}$$

$$L_o = \pi \left[\frac{1.0x7.7x10}{0.05}\right]^{\frac{1}{2}} \left[\ln \frac{1.16x0.8}{0.3}\right]^{-\frac{1}{2}}$$

= 116 m

Now determine C = D ln D/u = 7.7 ln 7.7/ πx .10

= 24.63 m. The value of C may be also estimated from **Figure 3.17** (which comes out to be 25 m). Compute L = L_o – 25 m = 91 m.

3.5.4 Discussion of the Glover-Dumm equation

Time averaged hydraulic head: For various reason, e.g. to account for the average horizontal flow above drain level, or to apply steady state equations, it may be required to compute a time averaged hydraulic head, h, between h_o and h_t or between h_{t1} and h_{t2} during tail recession.

During tail recession of hydraulic head, one could take the arithmetic mean of two readings (**Figure 3.39**).

$$\bar{h} = 1/2\,(h_o + h_t) \qquad 3.297$$

but then h will be over estimated because h_t changes according to an exponential function.

A better estimate of $\bar{h}$ is obtained by using geometric mean as follows :

$$\bar{h} = \sqrt{h_o h_t} \qquad 3.298$$

or

$$\log \bar{h} = \frac{1}{2}(\log h_o + \log h_t)$$

3.299

In the **Example 3.12** illustrating the Glover-Dumm equation, we would then get,

$\log \bar{h}$ =1 /2 (log 0.8 + log 0.3)= 1 /2 (-0.10- 0.52) = -0.31

So that $\bar{h}$ = 0.49 m. Now using the van Beers method, one will arrive at,

$$L_o = \sqrt{\frac{8KDh}{q}} = \sqrt{\frac{8x1.0x7.7x0.49}{0.0025}}$$

$$= 110 \text{ m}$$

C = D ln D/u = 7.7 ln 7.7/0.314 = 24.5

So, L = Lo – C = 110 – 24.5 m = 85.5 m

Which is slightly smaller than the spacing of 90 m which was computed in earlier calculation.

Alternatively, the average h may be defined as,

$$h = \frac{1}{t}\int_0^t h_t dt = \frac{1}{t}\int_0^t 1.16 h_o e^{-\alpha t} dt$$

3.300

which yields upon integration and rearrangement,

$$h = \frac{1.16h_o}{\alpha t}(1 - e^{-\alpha t}) = \frac{1.16h_o - h_t}{\ln(\frac{1.16h_0}{h_t})}$$

The flow above the drains should be taken into account if h is relatively large, or if D is small. Equation (3.278) then reads,

$$L = \sqrt{\left[\frac{\pi^2 K(d + \frac{1}{2}h)t}{f \ln 1.16 \frac{h_o}{h_t}}\right]}$$

3.301

Obviously large average $\bar{h}$ is needed in these cases. In most practical cases, however, this correction will have little or no effect on the drain spacing. **Figure 3.40** shows graphical solution of Glover-Dumm drain spacing equation.

3.5.5 Non-steady drain discharge

The discharge of the drains at time t, when expressed per unit surface area, can be found by applying Darcy's law.

$$q_t = \frac{-2KD}{L}\left[\frac{dh}{dx}\right]_{x=0}$$

3.302

Differentiating Equation (3.265) with respect to x, one will get,

$$\frac{dh}{dx} = -\frac{4}{\pi}h_o \sum_{n=1,3,5,..}^{\infty} \frac{1}{n}.\frac{n\pi}{L}e^{-n^2\alpha t}\cos\frac{n\pi x}{L}$$

3.303

$$\frac{dh}{dx} = -\frac{4}{\pi}h_o\frac{\pi}{L} \sum_{n=1,3,5,..}^{\infty} e^{-n^2\alpha t}\cos\frac{n\pi x}{L}$$

at x = 0,

$$\left|\frac{dh}{dx}\right|_{x=0} = -\frac{4}{L}h_o \sum_{n=1,3,5,..}^{\infty} e^{-n^2\alpha t}.1$$

3.304

Substituting this value of $|dh/dx|_{x=0}$ in Equation (3.302) one will obtain,

$$q_t = \frac{-2KD}{L}\left(-\frac{4h_o}{L}\right)\sum_{n=1,3,5,..}^{\infty} e^{-n^2\alpha t}$$

$$q_t = \frac{8KDh_o}{L^2}\sum_{n=1,3,5,..}^{\infty} e^{-n^2\alpha t}$$

3.305

Multiplying the denominator and numerator by $\pi^2 f$,

$$q_t = \frac{8\pi^2 f.KDh_o}{\pi^2 fL^2}\sum_{n=1,3,5,..}^{\infty} e^{-n^2\alpha t}$$

$$q_t = \frac{8}{\pi^2}\alpha.f.h_o \sum_{n=1,3,5,..}^{\infty} e^{-n^2\alpha t}$$

3.306

Ignoring higher terms of series one will get,

$$q_t = \frac{8}{\pi^2}.\alpha.f.h_o e^{-\alpha t}$$

3.307

Recalling the relationship between h_o and h_t

$$h_t = \frac{4}{\pi}h_o e^{-\alpha t} = 1.27 h_o e^{-\alpha t}$$

Combining Equation (3.269) and (3.307),

$$q_t = \frac{8}{\pi^2}.\alpha.f.\frac{\pi h_t}{4} = \frac{2}{\pi}.\alpha.f.h_t$$

3.308

Substituting the value of α, gives,

$$q_t = \frac{2}{\pi} \cdot \frac{\pi^2 Kd}{fL^2} \cdot f . h_t = \frac{2\pi Kd}{L^2} h_t$$

3.309

This is similar to Hooghoudt equation except that a factor of 2π is obtained in place of 8.

From Equation (3.269) and (3.307) it can be deduced that during tail recession,

$$\frac{q_{t2}}{q_{t1}} = \frac{h_{t2}}{h_{t1}} = e^{-\alpha(t2-t1)}$$

3.310

According to Equation (3.310), a plot of q_t or h_t on a logarithmic scale and time t on a linear scale will result a straight line. This relation is important. It can be used to determine the reaction factor, α, from field data of drained plots.

3.5.6 Integrated Hooghoudt drain spacing equation

Bouwer and van Schilfgaarde (1963) adopted steady state theory and abrupt drainage of pore spaces for predicting rate of fall of ground water in drained soil. They assumed that instantaneous drainage rate midway between the drains could be taken equal to the steady state drainage rate corresponding to the same water table height between drains. This was mathematically expressed as shown by Equation (3.285).

By substituting the value of q of Hooghoudt equation one will get.

$$\frac{4Kh^2 + 8Kdh}{L^2} = -fC\frac{dh}{dt}$$

3.311

$$Ah^2 + Bh = -fC\frac{dh}{dt}$$

3.312

where,

$A = 4K/L^2$

$B = 8Kd/L^2$

Equation (3.312) can be integrated for the limits, at t=0, h = h_o and at t=t, h = h_t.

$$\int_{h_0}^{h_t} \frac{dh}{Ah^2 + Bh} = \int_0^t dt$$

3.313

General solution of such type of equation is,

$$\int \frac{dx}{ax^2 + bx} = \frac{1}{b} \ln \frac{x}{ax + b}$$

3.314

Therefore after integration of Equation (3.313) one will get.

$$-\frac{1}{B} \ln \frac{h}{Ah + B}\Bigg|_{h_o}^{h_t} = \frac{t}{fC}\Bigg|_0^t$$

3.315

$$-\frac{1}{B} \ln \frac{h_t}{Ah_t + B} + \frac{1}{B} \ln \frac{h_0}{Ah_0 + B} = \frac{1}{fC}(t - 0)$$

$$\frac{1}{B} \ln \frac{h_0(Ah_t + B)}{h_t(Ah_0 + B)} = \frac{t}{fC}$$

$$\ln \frac{h_o(Ah_t + B)}{h_t(Ah_0 + B)} = \frac{Bt}{fC}$$

3.316

Substituting the value of A and B in above equation one will arrive at,

$$\ln \frac{h_o(\frac{4K}{L^2} h_t + \frac{8Kd}{L^2})}{h_t(\frac{4K}{L^2} h_0 + \frac{8Kd}{L^2})} = \frac{8Kdt}{fCL^2}$$

$$\ln \frac{h_o(\frac{4K}{L^2})(h_t + 2d)}{h_t(\frac{4K}{L^2})(h_0 + 2d)} = \frac{8Kdt}{fCL^2}$$

$$\ln \frac{h_o(h_t + 2d)}{h_t(h_o + 2d)} = \frac{8Kdt}{fCL^2}$$

$$L = \sqrt{\frac{8Kdt}{Cf \ln \frac{h_o(h_t + 2d)}{h_t(h_o + 2d)}}} \quad 3.317$$

Which is Integrated Hooghoudt drain spacing formula.

3.5.7 Integrated Toksöz -Kirkham drain spacing equation

Combining Equation (3.191) and (3.285) following equation may be obtained.

$$-fC\frac{dh}{dt} = \frac{K.h}{L.F_k.} \quad 3.318$$

by separating the variable one can also write,

$$-\frac{dh}{h} = \frac{K}{L.F_k.fC}dt \quad 3.319$$

Integration of Equation (3.319) under the limits as have been used in case Integrated Hooghoudt drain spacing formula.

$$-\int_{h_o}^{h_t} \frac{dh}{h} = \int_0^t \frac{K}{L.fC.F_k}dt$$

$$-\ln h\Big|_{h_o}^{h_t} = \frac{K}{L.fC.F_k} t\Big|_o^t$$

$$\ln \frac{h_o}{h_t} = \frac{Kt}{CfF_{K.}}$$

$$L = \frac{Kt}{C.f.F_k \ln \dfrac{h_o}{h_t}}$$

3.320

This is known as Integrated Toksoz Kirkham drian spacing formula. We need computer aid to compute factor F_k in order to utilize Integrated Toksoz drain spacing formula. Tabular values may be also used satisfactorily to estimate F_k. This equation may be easily employed for predicting fall of water table in tile drained land.

Example 3.13

Following observations were recorded from a waterlogged area in Mahi Right Bank Canal for designing of tile drainage system.

K=2.65 m

f = 0.14

D = 1.2 m

r_o = 0.10 m

Calculate the drain spacing using Integrated Hooghoudt drain spacing formula. Assume necessary data for calculation.

Solution 3.13

Assume that a 30 cm soil profile must be clear from water within 2 days and depth of drain to be laid in the area is 1.75 m below the soil surface. In rainy season when there is a continuous rain, water table may reach ground surface, therefore initial water table height may be taken as h_o=1.75 m for the worst situation.

Since 30 cm, soil profile has to be cleared from water within two days, therefore water table height above drain axis (hydraulic head), $h_t = 1.75$ m – 0.30 m = 1.45 m.

Now substituting all these values in Equation (3.317) will yield.

$$L = \left[\frac{8Kdt}{Cf \ln \dfrac{h_o(2d + h_t)}{h_t(2d + h_o)}} \right]^{\frac{1}{2}}$$

$$L = \left[\frac{8X\,2.65XdX\,2}{0.8X0.14X \ln \dfrac{1.75(2d + 1.45)}{1.45(2d + 1.75)}} \right]^{\frac{1}{2}}$$

$$L = \left[\frac{378.57d}{\ln \dfrac{1.21(2d + 1.45)}{(2d + 1.75)}} \right]^{\frac{1}{2}}$$

Above equation can be solved by trial and error.

Trial I: Assume a drain spacing of L= 70 m. Calculate the equivalent depth of impervious layer below drain axis, d with the help of following equation

$$d = \frac{D}{1 + \dfrac{8D}{\pi L} \ln \dfrac{D}{u}}$$

or using the van Beers nomograph or Hooghoudt table/nomograph or by using other relationships.

$$d = \frac{1.2}{1 + \frac{8x1.2}{\pi x70} \ln \frac{1.2}{\pi x0.10}} = 1.13m$$

Therefore,

$$L = \left[\frac{378.57x1.13}{\ln \frac{1.21(2x1.13 + 1.45)}{(2x1.13 + 1.75)}} \right]^{\frac{1}{2}} = 61.6m$$

Since calculated spacing is less than assumed spacing therefore one has to assume new spacing and perform calculation.

Trial II: Assume a new drain spacing L = 65 m and calculate or estimate the equivalent depth of impervious layer below drain axis by any of the method which comes out to be d=1.19 m (van Beer's equation).

$$d = \frac{1.2}{1 + \frac{8x1.2}{\pi x65} \ln \frac{1.2}{\pi x0.10}} = 1.13m$$

Calculate drain spacing using the earlier formula which comes out to be L = 61.6 m.

Calculated drain spacing is less than the assumed value, hence assume a new value of drain spacing and perform calculation as earlier.

Trial III: Assume drain spacing L= 60 m and estimate or calculate the equivalent depth of impervious layer below drain axis using any approach, which was calculated to be d=1.12 m from the van Beer's formula. Calculate drain spacing using the Integrated Hooghoudt drain spacing formula which was L= 61.6 m.

Calculated drain spacing is higher than the assumed value, hence increase the assumed value of drain spacing.

Trial IV: Assume drain spacing L=62 m. Calculate equivalent depth of impervious layer d from van Beer's formula which comes out to be d= 1.125 m. Calculate the drain spacing using Integrated Hooghoudt formula which is L= 61.5 m.

Assumed and calculated values of drain spacing are quite close and may be selected as final value.

Example 3.14

Following information were collected from a tile drained plot at Sampla village in Rohtak district of Haryana state near Delhi.

Drain spacing, L = 75 m

Saturated hydraulic conductivity, K = 4.10 m/day

Depth of impervious layer below drain axis, D = 1.2 m

Drainable porosity, f = 0.14

Radius of tile drains, r = 0.10 m

Depth of drains below ground surface = 1.75 m

After a rain midway water table reaches to a height of 1.60 m above drain axis. Using Integrated Toksöz Kirkham equation calculate the daily water table height for coming ten days period.

Solution 3.14

1. Calculate D/2r = 1.2/ (2x1.6) = 6, and L/D = 75/1.2 = 62.5.
2. Making use of these values estimate the value of F_K from **Table 3.3**, which comes out to be 8.21.
3. Substituting the soil hydrological parameters and value of F_K in Integrated Toksöz-Kirkham drain spacing formula as below :

$$75 = \frac{4.1t}{0.8x0.14X8.21X\ln\frac{1.60}{h_t}}$$

$$\ln\frac{1.60}{h_t} = \frac{4.1t}{0.8X0.14X8.21X75} = 0.0595t$$

$$\ln\frac{h_t}{1.60} = -0.0595t$$

$$h_t = 1.60e^{-0.0595t}$$

Now substituting the value of t in above equation,

t=1 day, h_1 = 1.51 m

t=2 day, h_1 = 1.42 m

t=3 day, h_1 = 1.34 m

t=4 day, h_1 = 1.26 m

t=5 day, h_1 = 1.19 m

t=6 day, h_1 = 1.12 m

t=7 day, h_1 = 1.06 m

t=8 day, h_1 = 0.99 m

t=9 day, h_1 = 0.94 m

t=10 day, h_1 = 0.88 m

3.5.8 van Schilfgaarde drain spacing equation

If the position of water table is permitted to vary, the following differential equation resulting from the Dupuit-Forchheimer assumptions can be used for two dimensional problems such as parallel drains. Equation reads as:

$$K\frac{\partial}{\partial x}(h\frac{\partial h}{\partial x}) + P = f.\frac{\partial h}{\partial t} \tag{3.321}$$

Because this nonlinear differential equation is difficult to solve, it is frequently linearized. One method of linearization is based on the assumption that the variation in h is small compared to $\partial h/\partial x$.

When there is no recharge equation (3.321) reduces to,

$$\frac{\partial}{\partial x}(h\frac{\partial h}{\partial x}) = \frac{f}{K}.\frac{\partial h}{\partial t} \qquad 3.322$$

When the drains do not rest on the impervious barrier, the van Schilfgaarde solve the Equation (3.322) and gave the following drain spacing formula.

$$L = 3A\left[\frac{K(d+h_t)(d+h_o)t}{2f(h_o-h_t)}\right]^{\frac{1}{2}} \qquad 3.323$$

Where 'A' was defined in terms of an incomplete beta- function. With negligible error, A can be represented by equation.

$$A^2 = \frac{h_o(2d+h_o)}{(d+h_o)^2} \qquad 3.324$$

van Schilfgaarde compared the performance of Equation (3.323) with Glover equation and showed that the differences in spacing as computed by the two equations were of the order of 10-40% for the range of practical interest. In the above equation correction was made for convergence by means of Hooghoudt' equivalent depth of impervious depth.

van Schilfgaarde (1964) also modified the Glover equation to correct for convergence near the drain and to take shape of water table into account by replacing the shape factor $4/\pi$ by a new variable shape factor $(2d_e+h_t)/(2d_e+h_o)$ as was obtained in Integrated Hooghoudt equation. After this change Glover equation becomes.

$$L = \sqrt{\frac{8Kdt}{Cf \ln(\frac{2d + h_t}{2d + h_o}) \frac{h_o}{h_t}}}$$

3.325

For most practical situation C = 0.9 and 8/0.8 ≅ 9, therefore Equation (3.325) reduces to,

$$L = \sqrt{\frac{9Kdt}{f \ln(\frac{2d + h_t}{2d + h_o}) \frac{h_o}{h_t}}}$$

3.326

van Schilfgaarde Equation (3.323) and Modified Glvoer Equation (3.326) can be used for drain spacing calculation by trial and error method.

References:

Aronovici, V.S. and W.W. Donnan. 1946. Soil permeability as a criterion for drainage design. Trans. Am. Geophys, Union.27:-101.

Boumans, J.H. 1963. Een algemene nomografische oplossing van het stationaire ontwateringsvraagstuk met toepassingen voor istroop en anirotroop doorlatende ontwateringsvraagustuk met toepassingen voor isotroop en anirotroop doorlatendegroden. Ploytechn. Tijdschr. 18(148): 545-551.

Bouwer, H. and J. Schilfgaarde. 1963. Simplified method of predicting fall of water table indrained land. Trans. of ASAE. 6(4): 288.

Carlslaw, H.S. and J.C. Jaeger. 1959. Conduction of heat in solids. Oxford University Press (2nd Edition) pp.510.

Colding, A. 1872. On the laws of water movement in soil. (In Danish) K.Dan, Videnski, Selsk, Skr, Naturvidensk. Mat. Afd. 9B8 : 563-621.

Dagan, G. 1964. Spacings of drains by an approximate method. J. Irrigation and Drainage Div. Proc. ASCE, Paper 3824, 41-46.

Donan, W.W.1946. Model tests of a tile spacing formual. Proc. Soil Science Soc. Of Am. 11:131:136.

Dumm, L.D.1954. Drain spacing formula. Agric. Engng. 35:726:730.

Dumm, L.D.1960. Validity and use of the transient flow concept in subsuface drainage. Paper presented at ASAE meeting, Memphis, Tenn. De c. 4-7.

Ernst, L.F.1956. Calculation of the steady state flow of ground water in vertical cross section. Neth. J. Agric. Science. 4:126-131.

Ernst, L.F.1962. Grondwaterstromingen in de verzaigde zone en hun berekening bij aanwezighied van horzontale evenwijdige open leidingen. Versl. Landbouwk. Onderz. 67-15, 189.

Hooghoudt, S.B.1936. Bepalingvan den doorlaatfaktor van den grond met behulp van pompproeven (z.g. boorgatenmethode). Versl. Landbouwk, Onderz. 42:449-541.

Hooghoudt, S.B.1937. Contributions to the knowledge of some physical soil properties. No. 6(in Dutch) VerslagenvanLandbouwkundige Onderzoekingenno. 43(13) B.Dep. van Economische Zaken Directe van den Landbouw. Algemeene Landsdrukkerij. The Haue.

Hooghoudt, S.B.1940. Bijdragen tot de kennis van enige natuurkundige groothedenvan deg grond. No. 7 Versl. Landbouwk, Onderz. 46:515-707.

ILRI.1974: Drainage Principles and Aplications, Volume II. Bull. No.16- Theories of field drainage and watershed ruoff, Wageningen, The Netherlands.

Kirkham, D.1958. Seepage of steady rainfall through soil into drains. Trans. Am. Geophys. Union 39:892-908.

Kirkham, D. 1960. An. Upper limit for the height of the water table in drainage design formulas. 7th International Congress of Soil Science. Madison I: 486-492.

Kozeny, J. 1931. About drain spacing (In German) Der Kulturtechniker, 34:226 : 232.

Kozeny, J. 1932. Hydrological principles of drain experiments (In Geman) In: Mtg. Commission VI ISS.Groningen, The Netherlands. July 1932, Vol. A Martinues Nijhoff, the hague, 42-67.

Labye, Y. 1960. Notre sur la formule de Hooghoudt. Bull. Tech. du Genie Rural, Min. de I' Agriculture de la Rep. Franciase, Paris No. 49-1, 21.

Rothe, J.1924. The drain spacing in mineral soils (In German). Der Kulturtechniker 32:155-169.

Smedma, L. K. and D.W. Rycroft. 1983. Land drainage: Planning and design of agricultural drainage systems. Batsford, London.

Toksoz, S. and D. Kirkham. 1961. Graphical solution and interpretation of a new drain spacing formula. J. Geophys. Res. 60:509-516.

Toksoz, S. and D. Kirkham. 1971. St eady drainage of layered soils: theory and nomographs. Journal of Irrigation and Drainage Division ASCE97(IR1):1-37.

US Bureau of Reclamation (USBR). 1978. Drainage Manual, Washington, D.C., U.S. Government Printing Office.

van Beers, W.F.J. 1979. Some nomographs for the calculation of drain spacings. 3rd edition. ILRI Bulletin 8, Wageningen.

van Beers, W.F.J. 1965. Some nomographs for the calculation of drain spacing.Bull. No.8. International Institute for Land Reclamation and Improvement. Wageningen.

van Schilfgaarde 1964. Design of tile drainage for falling water table. Journal of Irrigation and Drainage Division, American Society of Civil Engineers 90 (IR-3) : 71-73.

Wesseling, J.1964. A comparison of the steady state drain spacing formulas of Hooghoudt and Kirkham in connection with design pract ice. Journal fo Hydrology. 2:25-32.

Problems

Theoretical

1. Describe different types of drain with the help of a neat sketch and

their functioning.

2. Define uniform and non-uniform flow through porous medium.
3. What do you understand with the words steady state and transient or non-steady state?
4. State Depuit-Forchheimer assumptions used for simplification of ground water flow.
5. Make a definition sketch of ground water flow towards ditches and derive the ellipse equation for drain spacing calculation.
6. Derive Hooghoudt's steady state drain spacing formula and discuss about the terms of flow above drain level and flow below drain level.
7. Dervie an expresssion for maximum depth of flow considering resistance to vetical flow in Hooghoudt formula.
8. What do you understand with term equivalent depth of impervious layer below drains axis? Derive an expression for equivalent depth.
9. State the principle of Ernst steady state drain spacing formula and dervice the expression for total head required to flow towards drain in homogeneous soil.
10. Write the Kirkham's steady state drain spacing formula and its use for drain spacing calculation.
11. Write the Dagan's steady state drain spacing formula for both the conditions, when (1) drain is located at water table, and (2) drain is located below water table.
12. Derive Glover unsteady drain spacing formula for initially flat water table conditions. What is basic difference between Glover (1954) and Dumm (1960) drain spacing formula.
13. Dervice an unsteady state drain formula similar to Glover-Dumm equation by using Hooghoudt steady state drain spacing formula and Bouwer and Schilfgaarde (1963) approach, and compare the two formulas.
14. Derive integrated Hooghoudt drain spacing formula for transient conditions using Bouwer and Schilfgaarde approach.

15. Derive an expression for predicting transient drain discharge using Glover equation for hydraulic head variation with time.

16. Derive the Integrated Toksoz-Kirkham drain spacing formula for describing falling water table in absence of recharge.

Numerical

1. Calculate equivalent depth of impervious layer below drain axis if the drain spacing is 50m, depth to impervious layer from drain axis is 3.0 m and drain diameter is 0.20 m.

   [Anser: d = 2.30 m]

2. Calculate equivalent depth of impervious layer below drain axis using van Beer's generalized expression for equivalent depth, if the drain spacing is 80 m, depth to impervious layer from drain axis is 7.5 m and drain dimaeter 0.20 m.

   [Answer: d = 4.27 m]

3. From a water balance study of canal irrigated area requiring immediate drainage attention, the drainage coefficient was calcualted to be 7 mm per day and saturated hydraulic conductivity as 2.0 m per day. Depth of drain was decided to be 1.5 m below the ground surface. Calculate spacing between the drains with the help of Hooghoudt's formula, if the permisible water table height midway between the drains from ground surface is 1.2 m. Depth to impervious layer was measured to be 3.0 m below the drain axis.

   [Answer : 72.5m]

4. Following data were collected from a drainage investigatin for two layered soil.

   1. Saturated hydraulic conductivity above drain axis = 2.0 m/day
   2. Saturated hydraulic conductivity below drain axis = 1.0 m/day
   3. Depth to impervious layer below drain axis = 3 m
   4. Permisible water tale height above drain axis = 0.4m

   Calculate the spacing between the drains if the drain radius is

0.10 m. Assume necessary data required in calcualtion, if any.

[Answer: 33m]

5. For the given set of data, calculate spacing between drain using Kirkham's steady state formula and tabulated values of F_K.

   K= 0.8 m, q = 0.002 m/day, r = 0.10, D = 5.0 m and h = 0.60 m

   [Answer : 85 m]

   Hint: Assume some value of spacing and see the F_K- value from table and calculate drain spacing. Compare the calculated and assumed spacing, if they are equal or quite close to each other, the same may be adopted otherwise new drain spacing should be assumed and process is repeated.

6. Calculate spacing between the drains using Dagan's steady state drains spacing formula when drain is located at water level with the following set of data :

   K = 1.5 m/day, q=R = 20 mm per day, D = 3.5 m, h = 0.5 m and r = 0.10 m.

   [Answer : 24.5 m]

7. Calculate spacing between the drain using Ernst's steady state drain spacing formula for the following set of data in homogeneous soil.

   D_0=4 m, q = 0.002 m/day, r= 0.08, h = 0.75 m and K_1 = 1.0 m/day.

   [Answer : 104 m]

8. From a water balance study in a canal irrigated area it was found that after each irrigation 0.02 m depth of water need to removed within 10 days. Water level midway between the drains should never rise 1.0 m below the ground surface. Irrigation interval is 10 days, saturated hydraulic conductivity is 1.15 m/day and the effective drainable porosity is 0.05. If drains have to be laid down at a depth of 2.0 m below the ground surface, calculate spacing between the drains using Glover-Dumm non-steady state drains spacing formula. Depth to impervious layer is measured to be 7.5 m below the drainage level.

   [Answer : 133 m]

Chapter 4

WATER TABLE AND DRAIN DISCHARGE FLUCTUATION

Information on water table fluctuation is required to correlate the water table fluctuation with crop response in tile drained plot and efficacy of the designed drainage system. Optimization of drain spacing is also possible through the water table fluctuation models by predicting water table in response to rainfall or recharge or irrigation for several years. A large number of mathematical models are available for predicting water table in response of rainfall/irrigation. Some of the common models are being presented in this chapter.

4.1 de Zeeuw and Hellinga Model

de Zeeuw and Hellinga (1958, 1966) observed that storage in a polder area is mainly the ground water stored below the ground water table between the parallel tile drains or ditches. In their analysis of field data, they found an approximate proportionality of storage and outflow rate. They argued that a soil drainage system in such a polder area acts similarly as a linear reservoir. They took the simple Hooghoudt equation, that neglects flow above drain level.

$$q = \frac{8KDh}{L^2} = \frac{8KD}{L^2} h \tag{4.1}$$

We want to describe the non-steady state outflow, that is, the outflow as it varies with time, dq / dt,

$$\frac{dq}{dt} = \frac{8KD}{L^2} \cdot \frac{dh}{dt} \tag{4.2}$$

The water balance of a drained soil is roughly dependent on:

1. a rising water table, h, if R>q or (R-q)>0
2. a falling water table, h, if R<q, or (R-q)<0

As a result, the water table elevation as it varies with time, dh/dt, can be described as follows:

$$\frac{dh}{dt} = \frac{R-q}{Cf} \tag{4.3}$$

where,

C = a shape factor which takes into account the fact that the water table is not horizontal as it assumed in model concept, in this case, C≈0.8; and f is drainable porosity.

If we substitute Equation (4.3) in Equation (4.2) we will get,

$$\frac{dq}{dt} = \frac{8KD}{L^2}\frac{R-q}{Cf} = \frac{10KD}{fL^2}(R-q) = \alpha(R-q) \tag{4.4}$$

where,

α = the so called reaction factor, a constant which is determined by soil parameters K, D and f, and the drain spacing L, [1/T]

$$\alpha = \frac{10KD}{fL^2} \approx \frac{\pi^2 KD}{fL^2}$$

The reaction factor reflects the system response as follows:

1. a high value of the reaction factor [high KD, low f and small L]: h and q respond rapidly to changes in recharge;
2. a low value of the reaction (low KD, high f and large L): h and q respond slowly to changes in recharge.

de Zeeuw and Hellinga were the first to use such a compound hydrologic factor to characterize a drainage situation. The linear differential equation,

$$\frac{dq}{dt} = \alpha(R - q)$$

Describes drain outflow q with time as a function of a simple water balance. It is solved by integration for the following boundary conditions:

$$\begin{array}{ll} t = t; & q = q_t \\ t = t\text{-}1; & q = q_{t\text{-}1} \end{array} \Bigg|$$

(4.5)

as under,

$$\int_{t-1}^{t} \frac{dq}{dt} = \int_{t-1}^{t} \alpha dt$$

(4.6)

After integration one will obtain the following expression.

$$-\ln(R - q)\Big|_{q_{t-1}}^{q_t} = \alpha t\Big|_{t-1}^{t}$$

$$\ln\frac{(R-q)_t}{(R-q)_{t-1}} = -\alpha\{t-(t-1)\} = -\alpha\Delta t$$

$$\frac{(R-q)_t}{(R-q)_{t-1}} = e^{-\alpha\Delta t}$$

(4.7)

where,

$\Delta t = t - (t\text{-}1)$: the time interval between two readings, [T]

Rearranging Equation (4.7) one will obtain,

$$R_t - q_t = R_{t-1}e^{-\alpha\Delta t} - q_{t-1}e^{-\varepsilon\Delta t}$$

$$q_t = q_{t-1}e^{-\alpha\Delta t} + R_t - R_{t-1}e^{-\alpha\Delta t}$$

For constant recharge/replenishment, $R_t = R_{t-1} = R$. Denoting constant recharge over time interval Δt as $R_{\Delta t}$ above equation becomes.

$$q_t = q_{t-1} e^{-\alpha \Delta t} + R_{\Delta t}(1 - e^{-\alpha \Delta t}) \tag{4.8}$$

Alternatively, the q—h relationship as expressed in Equation (4.1) can be written as follows considering $\pi^2 \cong 10$.

$$q = \frac{8KD}{L^2} h = \frac{8KD}{L^2} \left(\frac{0.8f}{0.8f}\right) h = 0.8 \frac{10KD}{fL^2} fh = 0.8\alpha . f . h \tag{4.9}$$

Writing above equation in following forms :

$$q_t = 0.8 f\alpha . h_t \tag{4.10}$$

$$q_{t-1} = 0.8 f\alpha h_{t-1} \tag{4.11}$$

Substituting the latter two equations into Equation (4.8) will yields

$$0.8 f\alpha h_t = 0.8 f\alpha h_{t-1} e^{-\alpha \Delta t} + R_{\Delta t}(1 - e^{-\alpha \Delta t})$$

And finally one will obtain the desired model.

$$h_t = h_{t-1} e^{-\alpha \Delta t} + \frac{R_{\Delta t}}{0.8\alpha}(1 - e^{-\alpha \Delta t}) \tag{4.12}$$

This solution is appropriate to describe relatively slow variation of flow as is found in flat polder areas. Above equations can be satisfactorily used for predicting water table fluctuation in tile drained land when historical weather data and soil hydrological parameters are known. The non-steady processes are approximated by a series of interval during which recharge is assumed to be constant. The length of time interval should be such that changes occurring during interval may be described sufficiently accurately by the existing steady state formulas using as input average rates during the interval. This requirement is generally met by adapting interval of length $\Delta t = 1$ day. The first rain, after a dry period is normally mostly retained within the soil. In general, water table will not rise above the drainage base until the overlying soil is above field capacity. Consequently best simulations are obtained when these conditions exist e.g. during the latter period of a rain, lasting several days, or in moderate climates during winter when evapotranspiration is low and water tables are continuously high. In de Zeeuw Hellinga model depth of impervious layer below drain axis has to be replaced by equivalent depth of impervious layer to take radial flow into account.

4.2 Application of de Zeeuw-Hellinga model

The precipitation pattern, which is time dependent, is represented by individual rainstorms of constant intensity, with time, e.g. on a daily basis.

Example 4.1: The following data is available for a tile drained plot :

Drain spacing, L = 40 m

Depth of drain below ground surface = 1.2 m

Transmissivity, KD = 2.5 m^2/day

Drainable porosity, f = 0.05

Time interval, $\Delta t = 1$ day

For the given rainfall and evapotranspiration pattern predict daily water table height and drain discharge if initial water table height above drain level is 0.10 m and soil moisture deficit is 15 mm below field capacity one week after previous rain.

Day	0	1	2	3	4	5	6	7	8
Rainfall, mm	0	20	20	10	25	25	0	0	0
Evapotranspiration, mm	0	1	2	2	2	2	2	2	2

Solution 4.1

1. Calculate reaction factor from above data set as, $\alpha = (10KD)/(fL^2) = (10x2.5)/(0.05x40^2) = 0.31$.
2. Calculate $e^{-\alpha\Delta_t} = e^{-0.31x1} = e^{-0.31} = 0.73$.
3. Calculate $(1 - e^{-\alpha\Delta_t}) = 1 - 0.73 = 0.27$.
4. Calculate $0.8f\alpha = 0.8x0.05x0.31 = 0.12$
5. Substitute above calculated and given data in Equation (4.8) and (4.12), which will yield.

$$h_t = 0.73\ h_{t-1} + (0.27/0.012)R_{\Delta t}$$

$$q_t = 0.73\ q_{t-1} + 0.27R_{\Delta t}$$

These equations are employed to calculate water table height and drain discharge on daily basis in tabular form as given below :

day t	rainfall m R_t	evapotrans-piration, m E_t	moisture storage, m S_t	recharge m $R_{\Delta t}=R_t-E_t$	water table, m h_t	drain discharge m/day q_t
0	0.000	-	-0.15	-	0.10	0.001
1	0.020	0.001	-	0.004	0.16	0.002
2	0.020	0.001	-	0.019	0.54	0.007
3	0.010	0.002	-	0.008	0.57	0.007
4	0.025	0.002	-	0.023	0.96	0.012
5	0.005	0.002	-	0.003	0.77	0.010
6	0.000	0.002	-0.002	0.000	0.56	0.007
7	0.000	0.002	-0.004	0.000	0.41	0.005
8	0.000	0.002	-0.006	0.000	0.30	0.004

Calculation of daily water table height and drain discharge.

Water table and drain discharge fluctuations are presented in **Figure 4.1**.

4.3 Kraijenhoff van de Leur-Maasland model

Both **Kraijenhoff van de Leur (1958)** and **Maasland (1959)** independently derived the solution for non-steady state ground water flow to drains. The solution is based on a steady recharge over any time period,

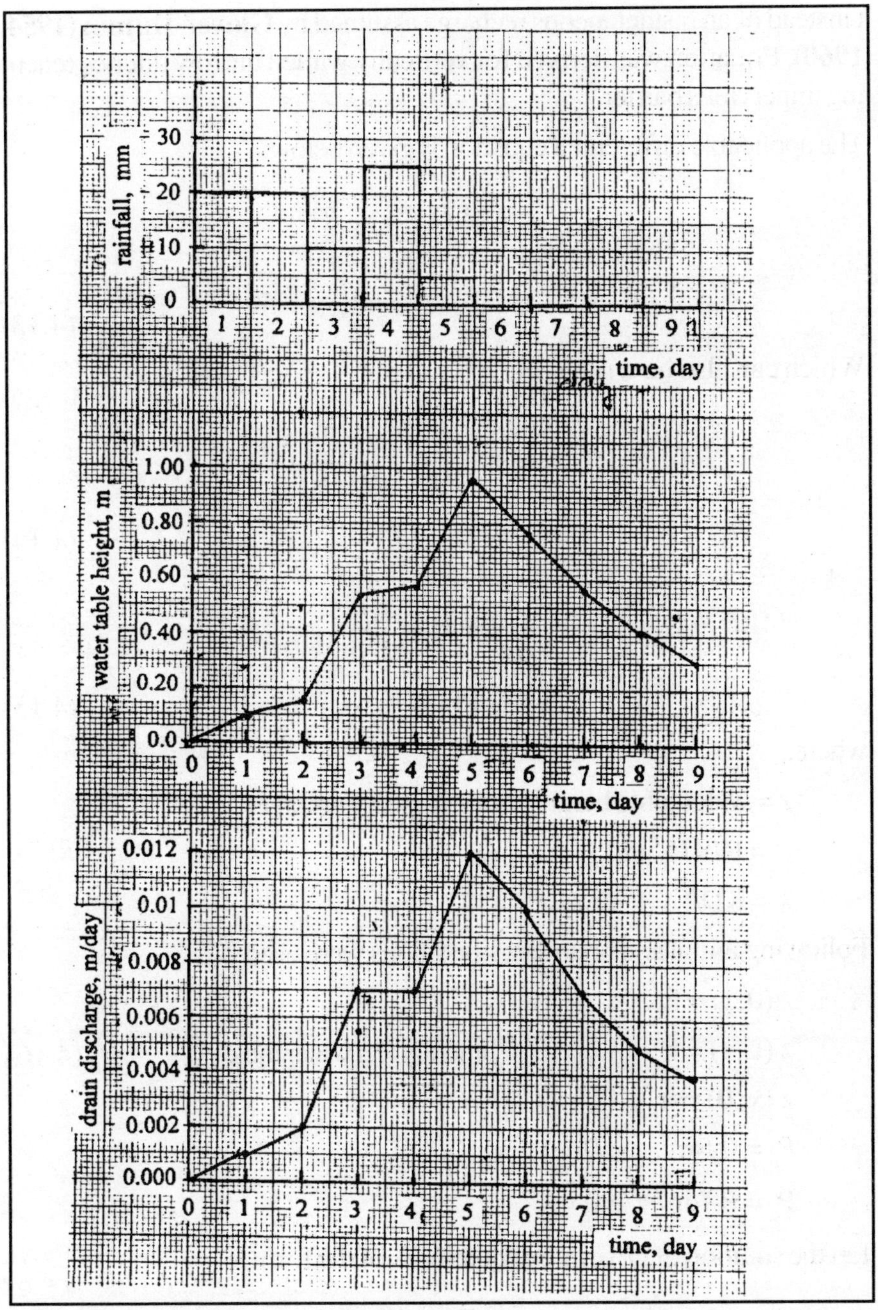

Figure 4.1. Graphical presentation of hydrological data, water table and discharge fluctuation.

t instead of an instantaneous recharge assumed by **Glover-Dumm (1954, 1960)**. Equation was derived for vertically walled narrow ditches reaching impervious barrier.

The applicable differential equation is given below:

$$\frac{\partial z}{\partial t} = \frac{KD}{f}\frac{\partial^2 z}{\partial x^2} + \frac{R}{f} \tag{4.13}$$

Which can also be written as,

$$\frac{\partial^2 z}{\partial x^2} + \frac{R}{KD} = \frac{f}{KD}\frac{\partial z}{\partial t} \tag{4.14}$$

$$\frac{\partial^2 z}{\partial x^2} + P_c = \frac{1}{k}\frac{\partial z}{\partial t} \tag{4.15}$$

where,

z = water table height above impervious layer, [L]

P_c = R/KD, and

k = KD/f

Following initial and boundary conditions were used.

$$\left.\begin{aligned} z(0,t) &= h_1 \\ z(L,t) &= h_1 \\ z(x,0) &= h_1 \end{aligned}\right\} \text{ initially flat water table} \tag{4.16}$$

$P_c = 0$ for $t < 0$

$P_c = R$ for $t \geq 0$

Let the solution of governing equation is of the form,

$$z = u + w \tag{4.17}$$

Where u is solution for steady state and w is for unsteady state conditions.

For obtaining solution for steady state conditions substitute the Equation (4.17) in Equation (4.15).

$$\frac{d^2u}{dx^2} = -p_c \tag{4.18}$$

Boundary conditions for steady state conditions (4.16) reduces to,

$$\left.\begin{aligned} u(0) &= h_1 \\ u(L) &= h_1 \end{aligned}\right] \tag{4.19}$$

Now from Equation (4.19)

$$\frac{du}{dx} = -P_c x + C_1 \tag{4.20}$$

$$u = -\frac{P_c}{2}x^2 + C_1 x + C_2 \tag{4.21}$$

Substitute the boundary conditions (4.19) in Equation (4.21), one will get.

$$h_1 = -\frac{P_c}{2} X 0^2 + C_1 X 0 + C_2 \Rightarrow C_2 = h_1$$

And,

$$h_1 = -\frac{P_c L^2}{2} + C_1 L + h_1 \Rightarrow C_1 = \frac{P_c L}{2}$$

Substituting the values of integration constant in Equation (4.21), the steady state solution becomes,

$$u = -\frac{P_c x^2}{2} + \frac{P_c L}{2} x + h_1 \Rightarrow h_1 - \frac{P_c x}{2}(x - L)$$

or

$$z = -\frac{P_c x^2}{2} + \frac{P_c L}{2} x + h_1 \Rightarrow h_1 - \frac{P_c x}{2}(x - L)$$

(4.22)

For unsteady state conditions Equation (4.17) can be substituted in Equation (4.15) which will yield.

$$\frac{\partial^2 w}{\partial x^2} = \frac{1}{k}\frac{\partial w}{\partial t}$$

(4.23)

Changed boundary conditions are given below :

$$\left.\begin{aligned} &w(0, t) = 0 \\ &w(L, t) = 0 \\ &w(x, 0) = f(x) - u \end{aligned}\right\} \quad (4.24)$$

at t = 0,

$$w = h_1 - \{h_1 - \frac{P_c x}{2}(x - L)\} = \frac{P_c x}{2}(x - L) = -\frac{P_c x}{L}(L - x)$$

(4.25)

Solution of Equation (4.23) under the boundary conditions (4.24) is as follows **(Carlslaw and Jaeger, 1959)** :

$$w = \sum_{n=1}^{\infty} a_n . \sin\frac{n\pi x}{L} e^{-(\frac{kn^2\pi^2}{L^2})t}$$

(4.26)

where,

a_n = is fourier series constant given as below :

$$a_n = \frac{2}{L}\int_0^L f(x)\sin\frac{n\pi x}{L}dx \tag{4.27}$$

Substituting Equation (4.25) into Equation (4.27) one will get,

$$a_n = \frac{2}{L}\int_0^L -\frac{P_c x}{2}(L-x)\sin\frac{n\pi x}{L}dx \tag{4.28}$$

Equation (4.28) can also be written as,

$$a_n = \frac{P_c}{L}\int_0^L (x^2\sin\frac{n\pi x}{L} - Lx\sin\frac{n\pi x}{L})dx$$

$$a_n = \frac{P_c}{2}\int_0^L x^2\sin\frac{n\pi x}{L}dx - \frac{P_c}{L}L\int_0^L x.\sin\frac{n\pi x}{L}dx \tag{4.29}$$

$$a_n = I_1 + I_2$$

Integrating first part of Equation (4.29) as below :

$$I_1 = \frac{P_c}{L}\int_0^L x^2\sin\frac{n\pi x}{L}dx$$

$$I_1 = \frac{P_c}{L}\left[\frac{2x}{\frac{n^2\pi^2}{L^2}}\sin\frac{n\pi x}{L} - \frac{x^2}{\frac{n\pi}{L}}\cos\frac{n\pi x}{L} + \frac{2}{\frac{n^3\pi^3}{L^3}}\cos\frac{n\pi x}{L}\right]_0^L \tag{4.30}$$

$$I_1 = \frac{P_c}{L}\left[\frac{2L}{\frac{n^2\pi^2}{L^2}}\sin\frac{n\pi L}{L} - \frac{L^2}{\frac{n\pi}{L}}\cos\frac{n\pi L}{L} + \frac{2}{\frac{n^3\pi^3}{L^3}}\cos\frac{n\pi L}{L} - \frac{2}{\frac{n^3\pi^3}{L^3}}\right] \tag{4.31}$$

$$I_1 = \frac{P_c}{L}\left[\frac{L^3}{n^2\pi^2}\sin n\pi - \frac{L^3}{n\pi}\cos n\pi + \frac{2L^3}{n^3\pi^3}\cos n\pi - \frac{2L^3}{n^3\pi^3}\right]$$

Since sine terms are always zero for any value of n=1,2,3,.., therefore Equation (4.31) can also be written as,

$$I_1 = \frac{P_c}{L}\left[-\frac{L^3}{n\,\pi}\cos n\pi + \frac{2L^3}{n^3\pi^3}\cos n\pi - \frac{2L^3}{n^3\pi^3}\right] \qquad (4.32)$$

Similarly integration of second term I_2 can be performed.

$$I_2 = -P_c\int_0^L x.\sin\frac{n\pi x}{L}dx$$

$$I_2 = -P_c\left[\frac{\sin\frac{n\pi x}{L}}{\frac{n^2\pi^2}{L^2}} - \frac{x\cos\frac{n\pi x}{L}}{\frac{n\pi}{L}}\right]_0^L$$

$$I_2 = -P_c\left[\frac{\sin\frac{n\pi}{L}}{\frac{n^2\pi^2}{L^2}} - \frac{L\cos\frac{n\pi L}{L}}{\frac{n\pi}{L}}\right] = P_c\left[-\frac{L^2}{n^2\pi^2}\sin n\pi + \frac{L^2\cos n\pi}{n\pi}\right] \qquad (4.33)$$

Since sine term is always zero for any n=1,2,3,…, therefore,

$$I_2 = \frac{P_c L^2\cos n\pi}{n\pi} \qquad (4.34)$$

Now the fourier constant can be worked out as,

$$a_n = I_1 + I_2 = P_c\left[-\frac{L^2\cos n\pi}{n\pi} + \frac{2L^2\cos n\pi}{n^3\pi^3} - \frac{2L^2}{n^3\pi^3}\right] + \frac{P_cL^2\cos n\pi}{n\pi}$$

$$a_n = PcL^2\left[-\frac{\cos n\pi}{n\pi} + \frac{2\cos n\pi}{n^3\pi^3} - \frac{2}{n^3\pi^3} - \frac{\cos n\pi}{n\pi}\right]$$

$$a_n = \frac{2P_cL^2}{n^3\pi^3}[\cos(n\pi) - 1]$$

(4.35)

when, n = 1, cos nπ - 1 = cos π - 1 = -1 – 1 = - 2

when, n = 2, cos nπ - 1 = cos 2 π - 1 = 1 – 1 = 0

when, n = 3, cos nπ - 1 = cos 3 π - 1 = - 1 – 1 = - 2

when, n = 4, cos nπ - 1 = cos 4 π - 1 = 1 - 1 = 0

Which means Equation (4.35) will have same value for only odd numbers.

$$\therefore a_n = \frac{2P_cL^2}{n^3\pi^3}[(-1)^3 - 1] = -\frac{4P_cL^2}{n^3\pi^3}$$

(4.36)

General solution of Equation (4.15) can be obtained by substituting the solution u and w in Equation (4.17) written as below:

$$z = u + w = [h_1 - \frac{P_cx}{2}(x-L)] - \sum_{n=1,3,5,..}^{\infty} -\frac{4P_cL^2}{n^3\pi^3}\sin\frac{n\pi x}{L}e^{-(\frac{kn^2\pi^2}{L^2})t}$$

(4.37)

$$z = h_1 - \frac{P_cx}{2}(x-L) - \frac{P_c}{2}(\frac{8L^2}{\pi^3})\sum_{n=1,3,5,..}^{\infty}\frac{1}{n^3}\sin\frac{n\pi x}{L}e^{-(\frac{n^2k\pi^2}{L^2})t}$$

Substituting the value of k = KD/f and P_c = P/KD in above equation.

$$z = h_1 - \frac{Px}{2KD}(x-L) - \frac{P}{2KD}(\frac{8L^2}{\pi^3}) \sum_{n=1,3,5,..}^{\infty} \frac{1}{n^3} \sin\frac{n\pi x}{L} e^{-(\frac{n^2 KD\pi^2}{fL^2})t}$$

(4.38)

For midway water table height, substitute x=(1/2)L, z = h_t and $\alpha = \pi^2KD/(fL^2)$ in above equation.

$$h_t = h_1 + \frac{PL^2}{8KD} - \frac{P}{2KD}(\frac{8L^2}{\pi^3}) \sum_{n=1,3,5,..}^{\infty} \frac{1}{n^3} \sin\frac{n\pi}{2} e^{-n^2\alpha t}$$

(4.39)

When water table at the drain is at zero level i.e. h_1=0, above equation reduces to,

$$h_t = \frac{PL^2}{8KD} - \frac{PL^2(4)}{KD\pi^3} \sum_{n=1,3,5,..}^{\infty} \frac{1}{n^3} e^{-n^2\alpha t}$$

$$h_t = \frac{4}{\pi}\frac{f}{f}\frac{PL^2}{KD\pi^2}[\frac{\pi^3}{32} - \sum_{n=1,3,5,..}^{\infty} \frac{1}{n^3} e^{-n^2\alpha t}]$$

Since $\pi^3/32 = \sum_{1}^{\infty} (1/n)^3$ above equation takes the following form:

$$h_t = \frac{4}{\pi}\frac{P}{f} \cdot j \sum_{n=1,3,5,..}^{\infty} \frac{1}{n^3}(1 - e^{-\frac{n^2 t}{j}})$$

(4.40)

where,

j = 1/α, reservoir coefficient, since it represents the total storage in the aquifer, [T]

After sufficiently long time (t→∞) Equation (4.40) reduces to the following form:

$$h = \frac{4}{\pi}\frac{P}{f}.j\sum_{n=1,3,5,..}^{\infty}\frac{1}{n^3}(1) = \frac{4}{\pi}\frac{P}{f}\frac{fL^2}{\pi^2 KD}.\frac{\pi^3}{32} = \frac{PL^2}{8KD} = \frac{qL^2}{8KD}$$

(4.41)

This is actually Hooghoudt steady state equation when flow above drain level is ignored.

4.4 Drain discharge

The discharge of the drains at time t, when expressed per unit surface area, can be found by applying Darcy's law.

$$q_t = \frac{-2KD}{L}\left[\frac{dh}{dx}\right]_{x=0}$$

(4.42)

Value of h from Equation (4.38) can be substituted in above equation.

$$q_t = -\frac{2KD}{L}\frac{d}{dx}[-\frac{Lx - x^2}{2} - \frac{4L^2}{\pi^3}\sum_{n=1,3,5,..}^{\infty}\frac{1}{n^3}\sin\frac{n\pi x}{L}e^{-n^2\alpha t}]\frac{P}{KD}$$

(4.43)

$$q_t = -\frac{2KD}{L}[-\frac{L-2x}{2} - \frac{4L^2}{\pi^3}\sum_{n=1,3,5,..}^{\infty}\frac{1}{n^3}\cos\frac{n\pi x}{L}(\frac{n\pi}{L})e^{-n^2\alpha t}]\frac{P}{KD}$$

At x = 0

$$q_t = -\frac{2KD}{L}[-\frac{L}{2} + \frac{4L^2}{\pi^3}\frac{\pi}{L}\sum_{n=1,3,5,..}^{\infty}\frac{1}{n^3}n.1.e^{-n^2\alpha t}]\frac{P}{KD}$$

$$q_t = -\frac{2KD}{L}[-\frac{L}{2} + \frac{4L}{\pi^2}\sum_{n=1,3,5,..}^{\infty}\frac{1}{n^2}.e^{-n^2\alpha t}]\frac{P}{KD}$$

$$q_t = -\frac{8KD}{\pi^2}[-\frac{\pi^2}{8} + \sum_{n=1,3,5,..}^{\infty}\frac{1}{n^2}.e^{-n^2\alpha t}]\frac{P}{KD}$$

Since $\pi^3/8 = \sum_{1}^{\infty} (1/n^2)$

$$q_t = -\frac{8KD}{\pi^2}\frac{P}{KD}\left[\sum_{n=1,3,5}^{\infty}\frac{1}{n^2} + \sum_{n=1,3,5,..}^{\infty}\frac{1}{n^2}.e^{-n^2\alpha t}\right]$$

$$q_t = -\frac{8P}{\pi^2}\sum_{n=1,3,5}^{\infty}\frac{1}{n^2}(1-e^{-n^2\alpha t})$$

(4.44)

Above equations are valid as long as constant recharge rate P continues. When such a recharge occurs long enough, the flow conditions must become steady state too. For $t \to \infty$

Equation (4.44) reduces to,

$$q_t = -\frac{8P}{\pi^2}\sum_{n=1,3,5,..}^{\infty}\frac{1}{n^2}. = \frac{8P}{\pi^2}.\frac{\pi^2}{8} = P \qquad (4.45)$$

Which gives steady state condition where the discharge intensity q equals the recharge intensity.

When introducing the equivalent depth to impervious layer, d instead of D, to account for the convergence of streamlines in the vicinity of drains not reaching an impermeable layer, Equation (4.40) and Equation (4.44) remain same.

4.5 Application of Kraijenhoff van de Leur Maasland model

The Kraijenhoff van de Leur- Maasland model is not used for routine drain spacing calculations, which are usually based on an assumed steady or instantaneous recharge. The equations, however, prove very useful,

- when changes in water table elevation, h, and discharge rate, q, must be known for chosen drainage condition, and
- in response to a changing recharge pattern.

Such types of calculations are usually computerized. In the present section Kraijenhoff van de Leur- Maasland equation will be applied in order of increasing complexity, viz.

1. Constant and continuous recharge.
2. Constant recharge during a restricted period.
3. Intermittent recharge.

4.5.1 Constant and continuous recharge

Equation (4.40) and (4.44) can be written in abbreviated form for convenience, as under:

$$h_t = \frac{P}{f} jc_t \qquad (4.46)$$

Where,

$$c_t = \frac{4}{\pi} \sum_{n=1,-3,5}^{\infty} \frac{1}{n^3}(1-e^{-\frac{n^2 t}{j}}) \qquad (4.47)$$

And similarly,

$$q_t = Pg_t \qquad (4.48)$$

Where,

$$g_t = \frac{8}{\pi^2} \sum_{n=1,-3,5}^{\infty} \frac{1}{n^2}(1-e^{-\frac{n^2 t}{j}}) \qquad (4.49)$$

The factor c_t and g_t are depending on the time, t, and the reservoir coefficient, j, are tabulated in **Table 4.1**.

Example 4.2

Assume a drainage system with j = 5 day. The soil has an effective porosity f = 0.04. There is a constant recharge of 10 mm/day. Calculate water table and drain discharge fluctuation.

Table 4.1 Coefficients (c_t and g_t) for theKrainjenhoff van de Leur-Maasland equation.

t/j	g_t	c_t	t/j	g_t	c_t	t/j	c_t	c_t
0.01	0.072	0.010	0.48	0.497	0.447	1.0	0.730	0.809
0.02	0.102	0.020	0.50	0.57	0.463	1.15	0.743	0.830
0.03	0.125	0.030	0.52	0.518	0.477	1.20	0.756	0.850
0.04	0.143	0.0.39	0.54	0.528	0.492	1.25	0.767	0.869
0.05	0.161	0.049	0.56	0.537	0.507	1.30	0.779	0.887
0.06	0.176	0.060	0.58	0.546	0.521	1.35	0790	0.903
0.07	0.190	0.070	0.60	0.554	0.535	1.40	0.800	0.920
0.08	0.203	0.080	0.62	0.563	0.549	1.45	0.810	0.935
0.09	0.215	0.090	0.64	0.572	0.563	1.50	0.819	0.950
0.10	0.227	0.100	0.66	0.580	0.576	1.55	0.828	0.964
0.12	0.249	0.10	0.68	0.588	0.588	1.60	0.836	0.977
0.14	0.269	0.139	0.70	0.597	0.602	1.65	0.844	0.989
0.16	0.288	0.159	0.72	0.605	0.614	1.70	0.852	1.002
0.18	0.305	0.179	0.74	0.612	0.627	1.75	0.859	1.012
0.20	0.321	0.1992	0.76	0.620	0.638	1.80	0.866	1.023
0.22	0.337	0.218	0.78	0.628	0.650	1.85	0.872	1.033
0.24	0.352	0.238	0.80	0.636	0.661	1.90	0.879	1.033
0.26	0.367	0.257	0.82	0.643	0.672	1.95	0.885	1.052
0.28	0.380	0.275	0.84	0.650	0.683	2.00	0.990	1.061
0.30	0.373	0.294	0.86	0.657	0.675	2.10	0..901	1.078
0.32	0.406	0.312	0.88	0.663	0.706	2.20	0.910	1.093
0.34	0.419	0.329	0.90	0.670	0.717	2.30	0.919	1.107
0.36	0.430	0.347	0.92	0.677	0.727	2.40	0.927	1.18
0.38	0.442	0.364	0.94	0.683	0.737	3.00	0.960	1.71
0.40	0.454	0.381	0.96	0.689	0.746	4.00	0.985	1.210
0.42	0.465	0.398	0.98	0.696	0.756	5.00	0.995	1.226
0.44	0.476	0.415	1.00	0.202	0.765	∞	1.000	π^2 =R
0.46	0.487	0.461	1.05	0.715	0.787			1.232

Solution 4.2

Select values of g_t and c_t from Table 4.1 and perform calculation as given below.

Time	c_t	g_t	t/j	h_t=(P/f)jc_t m	q_t= R.g_t m/day
04 hour = 1/6 day	0.003	0.131	0.033	0.041	0.00131
08 hour = 1/3 day	0.067	0.184	0.067	0.084	0.00184
12 hour = 1 /2 day	0.100	0.227	0.100	0.125	0.00227
16 hou r = 2/3 day	0.133	0.262	0.133	0.166	0.00262
20 hour = 5/6 day	0.166	0.292	0.166	0.208	0.00292
24 hour = 1 day	0.199	0.321	0.200	0.249	0.00321
48 hour = 2 day	0.381	0.454	0.400	0.476	0.00454
72 hour = 3 day	0.535	0.554	0.600	0.669	0.00554
96 hour = 4 day	0.661	0.636	0.800	0.827	0.00636
120 hour = 5 day	0.765	0.702	1.000	0.956	0.00702
∞	π/8=1.232	1.000	∞	1.540	0.01000

4.5.2 Constant recharge during a restricted period

Consider a drained area with irrigation or rainfall, occurring during one single day, followed by a dry period. In order to compute water table heights on the days subsequent to the irrigation or rainfall, we assume (**Figure 4.2**) that the recharge P of the first day continues throughout the following days, but, simultaneously, from the second day onwards, an equal negative recharge, -P, is taken into account so that the total recharge is equal to zero from the second day onwards.

According to Equation (4.46), the water table height at the end of the first day (1 day) is given by,

$$h = h_1 = \frac{P}{f} jc_1$$

At the end of second day we have had a positive recharge P over two days, hence.

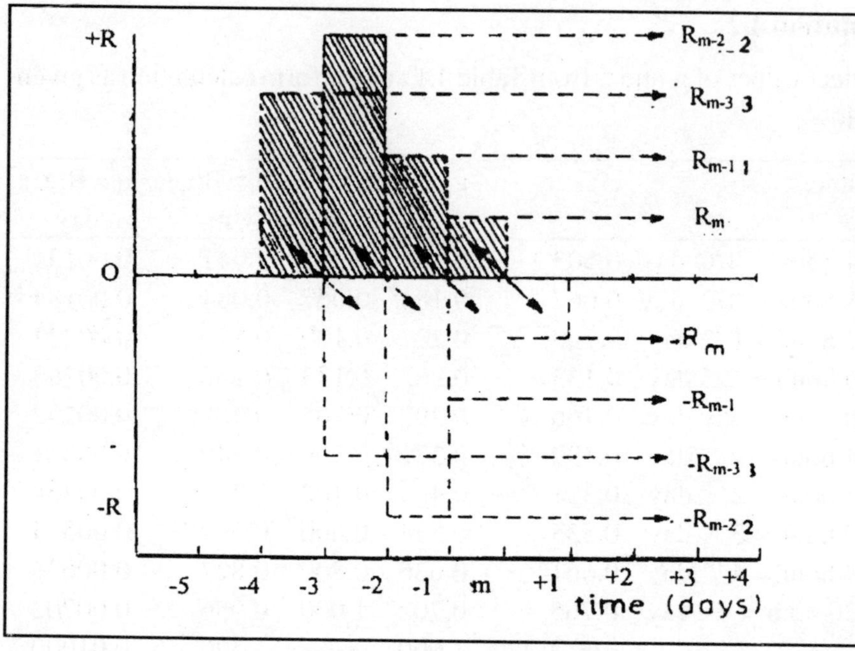

Figure 4.2. Principle of superpositions of rainfall/recharge P and water table h_t elevation.

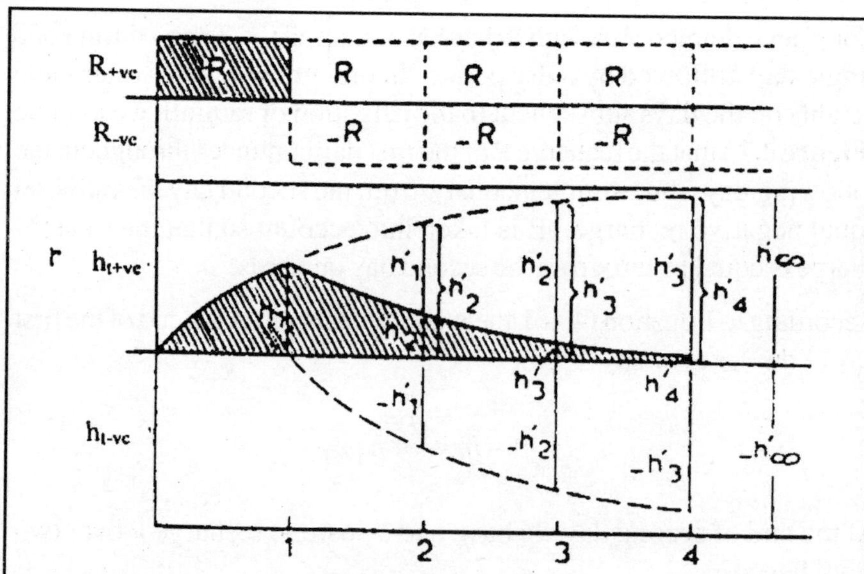

Figure 4.3. Superposition of intermittent rainfall/recharge.

$$h_2' = \frac{R}{f}\, j.c_2$$

From which we have to subtract the effect of a negative recharge over 1 day, equalling.

$$h_1' = \frac{R}{f}.j.C_1$$

So that,

$$h_2 = h_2' - h_1' = \frac{R}{f}\, j.(c_2 - c_1)$$

Similarly at the end of third day, we have a positive recharge P over three days, hence

$$h_3' = \frac{P}{f}.j.c_3$$

From which we have to subtract the effect of negative recharge over two days, i.e.

$$h_2' = \frac{P}{f}.j.c_2$$

So that,

$$h_3 = h_3' - h_2' = \frac{P}{f}\, j.(c_3 - c_2)$$

And at the end of the day,

$$h_t = h_t' - h_{t-1}' = \frac{P}{f}.j.(c_t - c_{t-1}) \tag{4.50}$$

The height of water table during the recession period can thus be computed with the aid of **Table 4.1**.

Example 4.3

Consider an area with tile drainage system at a depth of 1.0 m below soil surface, and the impermeable layer at a depth of 1.2 m below the drains as shown. The drain diameter is 0.20 m and the drain spacing is 20 m, so that equivalent depth to impervious layer d=1.0 m. The hydraulic conductivity of the soil is 0.5 m per day and effective drainable porosity is 0.05. Initial water table is at the drain level. During the first day a total amount of 30 mm of percolation water reaches the ground water due to an irrigation or rain. There was no percolation during following days. What will be the height of the water table midway between the drains during the days subsequent to the irrigation or rain.

Solution 4.3

Following parameters are given,

$K = 0.5$ m/day

$f = 0.05$

$D = 1.2$ m

$d = 1.0$ m

$L = 20$ m

$r_o = 0.10$ m

1. Calculate reservoir coefficient, $j = 1/\alpha$ as below.

$$j = \frac{fL^2}{\pi^2 KD} = \frac{0.05X\,20^2}{\pi^2 X0.50X1.0} = 4day$$

2. Select the value of c_t from Table 4.1 and perform calculation as given below:

time, t	t/j	c_t	c_{t-1}	c_t-c_{t-1}	(P/f)j	h_t = (P/f)j.(c_t-c_{t-1})
1 day	0.25	0.248	0.000	0.248	2.4	0.60 m
2 day	0.50	0.463	0.248	0.215	2.4	0.52 m
3 day	0.75	0.631	0.463	0.168	2.4	0.40 m
4 day	1.00	0.765	0.631	0.134	2.4	0.32 m

4.5.3 Intermittent recharge

The above method can be worked out in a more generalized way for intermittent recharge. Hydrological data are generally available on daily basis, therefore calculation should be done on daily basis; however, procedure holds good for any time period.

If one wishes to compute the height of water table or drain discharge at the end of any arbitrary day, let's choose it m[th] day (refer **Figure 4.3**). Both the height of water table and discharge rate is influenced by percolation during each of the preceding days. So one has to take into account;

- the recharge R_m over 1 day
- the recharge R_{m-1} over 2 days minus the recharge R_{m-1} over one day
- the recharge R_{m-3} over 3 days minus the recharge R_{m-2} over two days, and so on.

The height of the water table is thus given by,

$$h_m = \frac{j}{f}[R_m c_1 + R_{m-1}(c_2 - c_1) + R_{m-2}(c_3 - c_2) + ... + R_1(c_m - c_{m-1})]$$

(4.51)

Putting $C_1 = c_1 j$, $C_2 = (c_2\text{-}c_1)j$, $C_3 = (c_3\text{-}c_2).j$, and $C_m = (c_m - c_{m-1}).j$, one will obtain.

$$h_m = \frac{j}{f}[C_1 R_m + C_2 R_{m-1} + C_3 R_{m-2} + ... + C_m R_1]$$

(4.52)

Similarly the discharge rate is given by,

$$q_m = [G_1 R_m + G_2 R_{m-1} + G_3 R_{m-2} + ... + G_m R_1]$$

(4.53)

Where, $G_1=g_1$, $G_2=g_2\text{-}g_1$, $G_3 = g_3\text{-}g_2$, ... , $G_m = (g_m\text{-}g_{m-1})$

The values of C_t and G_t are given in **Table 4.2** and **Table 4.3**.

Table 4.2 Values of $C_t \times 10^2$ for the computation of transient water levels with Kraijenhoff van de Leur-Maasland equation.

? / T	0.01	.02	.03	.04	.05	.06	.07	.08	.09	.10	.11	.12	.13	.14	.15	.16	.17	.18	.19	.20
1	100	100	100	100	100	100	100	100	100	100	100	100	100	100	100	100	100	100	100	99
2	100	100	100	100	100	100	100	100	100	99	99	98	97	97	96	95	94	93	92	91
3	100	100	100	100	100	99	98	97	96	95	93	91	90	88	86	84	82	80	79	77
4	100	100	100	100	98	97	95	93	90	88	85	83	80	78	75	72	70	68	65	63
5	100	100	100	98	96	93	90	87	84	80	77	74	71	68	65	62	59	57	54	52
6	100	100	99	96	93	89	85	81	77	73	69	66	62	59	56	53	50	47	45	42
7	100	100	98	94	90	85	80	75	71	66	62	58	55	51	48	45	42	40	37	35
8	100	100	93	91	86	80	75	70	65	60	56	52	48	45	41	38	36	33	31	28
9	100	99	94	89	82	76	70	64	59	54	50	46	42	39	36	33	30	28	25	23
10	100	98	92	86	79	72	65	59	54	49	45	41	37	34	31	28	25	23	21	19
11	100	97	90	83	75	68	61	55	50	45	40	36	33	29	26	24	21	19	17	16
12	100	96	88	80	71	64	57	51	45	40	36	32	29	25	23	20	18	16	14	13
13	100	95	86	77	68	60	53	47	41	37	32	28	25	22	20	17	15	13	12	11
14	99	93	84	74	65	57	49	43	38	33	29	25	22	19	17	15	13	11	10	9
15	99	92	81	71	62	53	46	40	35	30	26	22	19	17	14	12	11	9	8	7
16	99	91	79	68	59	50	43	37	32	27	23	20	17	15	12	11	9	8	7	6
17	99	89	77	66	56	47	40	34	29	24	21	18	15	13	11	9	8	7	6	5
18	98	88	75	63	63	44	37	31	26	22	19	16	13	11	9	8	6	5	5	4
19	98	86	73	61	50	42	35	29	24	20	17	14	12	10	8	7	5	5	4	3
20	98	85	71	58	48	40	33	27	22	18	15	12	10	8	7	6	5	4	3	3

? / T	.21	.22	.23	.24	.25	.26	.27	.28	.29	.30	.31	.32	.33	.34	.35	.36	..37	.38	.39	.40
1	99	99	99	99	99	99	98	98	98	98	98	97	97	97	97	96	96	96	96	95
2	90	89	88	87	86	85	84	83	82	80	79	78	77	76	75	74	73	72	71	70
3	75	73	72	70	68	66	65	63	62	60	59	57	56	55	53	52	51	50	48	47
4	61	59	57	55	53	51	50	48	46	45	43	42	40	39	38	36	35	34	33	32
5	50	47	45	43	41	40	38	36	35	33	32	30	29	28	26	25	24	23	22	21
6	4	38	36	34	32	31	29	27	26	25	23	22	21	20	19	18	17	16	15	14
7	33	31	29	27	25	24	22	2	19	18	17	16	15	14	13	12	12	11	10	10
8	26	24	23	21	20	18	17	16	15	13	12	12	11	10	9	9	8	7	7	6
9	21	20	18	17	15	14	13	12	11	10	9	8	8	7	7	6	6	5	5	4
10	17	16	14	13	12	11	10	9	8	7	7	6	6	5	5	4	4	3	3	3
11	14	13	11	10	9	8	7	7	6	5	5	4	4	4	3	3	3	2	2	2
12	11	10	9	8	7	6	6	5	5	4	4	3	3	3	3	2	2	2	1	1
13	9	8	7	6	6	5	4	4	3	3	3	2	2	2	2					
14	7	7	6	5	4	4	3	3	3	2	2	2	1	1						
15	6	5	5	4	3	3	3	2	2	2	1	1								
16	5	4	4	3	3	2	2	2	1	1										
17	4	3	3	2	2	2	1	1												
18	3	3	2	2	2	1														
19	3	2	2	1	1															
20	2	2	1	1																

? / T	.41	.42	.43	.44	.45	.46	.47	.48	49	.50	.51	.52	.53	.54	.55	.56	5.7	.58	.59	.60
1	95	95	95	94	94	94	93	93	93	92	92	92	91	91	91	90	90	90	90	89
2	69	68	67	66	65	64	63	62	62	61	60	59	58	57	56	56	55	54	53	53
3	46	45	44	43	42	41	40	39	38	37	36	35	34	33	33	32	31	30	30	29
4	31	29	28	28	27	26	25	24	23	22	22	21	20	20	19	18	18	17	16	16
5	20	19	19	18	17	1	16	15	14	14	13	12	12	11	11	10	10	9	9	9
6	13	13	12	11	11	10	10	9	9	8	8	7	7	7	6	6	6	5	5	5
7	9	8	8	7	7	6	6	6	5	5	5	4	4	4	4	3	3	3	3	3
8	6	5	5	5	4	4	4	4	3	3	3	3	2	2	2	2	2	2	2	1
9	4	4	3	3	3	3	2	2	2	2	2	2	1							
10	3	2	2	2	2	2	1	1												
11	2	2	1																	

Table 4.3 Values of $G_t \times 10^3$ for the computation of transient drain discharge with Kraijenhoff van de Leur-Maasland equation.

? T	0.01	.02	.03	.04	.05	.06	.07	.08	.09	.10	.11	.12	.13	.14	.15	.16	.17	.18	.19	.20
1	72	102	124	144	161	176	190	203	216	227	238	249	259	269	278	287	296	305	313	321
2	30	47	52	60	66	73	79	84	89	94	99	103	107	111	115	119	123	126	130	133
3	23	32	40	46	51	56	60	64	68	72	76	79	82	85	88	91	94	96	98	100
4	19	27	33	38	43	47	51	54	58	61	64	66	69	71	73	75	77	78	80	81
5	17	24	29	34	38	42	45	48	51	53	55	58	60	61	62	63	64	65	66	66
6	15	22	26	31	34	38	40	43	45	47	49	5	52	53	53	54	54	54	54	54
7	14	20	24	28	31	34	37	39	41	42	44	45	45	46	46	46	46	45	45	44
8	13	18	23	26	29	32	34	36	37	38	39	40	40	40	40	39	38	38	37	36
9	12	17	21	25	27	30	32	33	34	35	35	35	35	34	34	33	32	32	31	30
10	12	16	20	23	26	28	29	30	31	31	31	31	31	30	29	28	27	26	25	24
11	11	16	19	22	24	26	27	28	28	28	28	28	27	26	25	24	23	22	21	20
12	10	15	18	21	23	24	25	26	26	26	25	4	24	23	22	21	20	18	17	16
13	10	14	18	20	22	23	24	24	24	23	22	22	21	20	19	18	16	15	14	13
14	10	14	17	19	21	22	22	22	22	21	20	19	18	17	16	15	14	13	12	11
15	9	13	16	18	20	20	21	20	20	19	18	17	16	15	14	13	12	11	10	9
16	9	13	16	18	19	19	19	19	18	17	16	15	14	13	12	11	10	9	8	7
17	9	12	15	17	18	18	18	17	16	15	14	13	12	11	10	9	8	7	7	6
18	8	12	15	16	17	17	17	16	15	14	13	12	11	10	9	8	7	6	6	5
19	8	12	14	16	16	16	16	15	14	13	12	10	10	8	8	7	6	5	5	4
20	8	11	14	15	15	15	14	14	13	12	10	9	8	7	7	6	5	4	4	3

? T	.21	.22	.23	.24	.25	.26	.27	.28	.29	.30	.31	.32	.33	.34	.35	.36	..37	.38	.39	.40
1	329	337	344	352	359	366	373	380	387	393	400	406	413	419	425	431	437	443	448	454
2	136	139	142	145	148	151	154	156	159	161	164	166	168	170	172	174	176	178	180	182
3	103	104	106	108	110	111	112	114	115	116	117	118	118	119	119	120	120	120	120	120
4	82	83	84	84	85	85	85	86	86	85	85	85	85	84	84	83	83	82	81	80
5	66	66	66	66	66	66	65	64	64	63	62	62	61	60	59	58	57	56	55	54
6	54	53	53	52	51	50	50	49	48	47	46	45	44	43	42	40	39	38	37	36
7	44	43	42	41	40	39	38	37	36	35	34	32	31	30	29	28	27	26	25	24
8	35	34	33	32	31	30	29	28	27	26	25	24	23	22	21	20	19	18	17	16
9	29	28	26	25	24	23	22	21	20	19	18	17	16	15	14	14	13	12	12	11
10	23	22	21	20	19	18	17	16	15	14	13	12	12	11	10	10	9	8	8	7
11	19	18	17	16	15	14	13	12	11	10	10	9	8	8	7	7	6	6	5	5
12	15	14	13	12	11	11	10	9	8	8	7	6	6	6	5	5	4	4	3	3
13	12	11	10	10	9	8	8	7	6	6	5	5	4	4	3	3				
14	10	9	8	8	7	6	6	5	5	4	4	3	3	3						
15	8	7	7	6	5	5	4	4	4	3	3	2								
16	6	6	5	5	4	4	3	3	3	2										
17	5	5	4	4	3	3	2	2												
18	4	4	3	3	2	2														
19	4	3	3	2	2															
20	3	2	2	2																

? T	.41	.42	.43	.44	.45	.46	.47	.48	.49	.50	.51	.52	.53	.54	.55	.56	5.7	.58	.59	.60
1	460	465	471	476	482	487	492	497	502	57	512	517	522	527	532	536	541	546	550	555
2	183	185	186	188	189	190	191	192	193	194	195	196	197	198	198	199	200	200	201	201
3	120	120	120	120	119	119	119	118	118	117	117	116	116	115	114	113	113	112	111	110
4	80	79	78	77	76	75	74	73	72	71	70	69	68	67	66	65	64	63	62	60
5	53	52	51	50	48	47	46	45	44	43	42	41	40	39	38	37	36	35	34	33
6	35	34	33	32	31	30	29	28	27	26	25	24	24	23	22	21	20	20	19	18
7	23	22	21	20	20	19	18	17	17	16	15	14	14	13	13	12	12	11	10	10
8	15	15	14	13	12	12	11	11	10	10	9	9	8	8	7	7	6	6	6	5
9	10	10	9	8	8	8	7	7	6	6	5	5	5							
10	7	6	6	5	5	5	4	4												
11	4	4	4																	

Example 4.4

A tile drainage system with a reaction factor of 0.25 day^{-1} [j = 4 day] in a soil with an effective drainable porosity of 0.05 receives the following rainfall :

July	15	16	17	18	19	20
Rainfall (mm)	5	20	10	5	0	0

What will be water table height above drain axis and drain discharge rate if initial water table is horizontal on 14th July, and at drain depth?

Solution 4.4

1. Consider successively July 20, 19, 18, 17, 16, and 15 as m[th] day and arrange them as shown in table given below.
2. Select C_t values from **Table 4.2** and arrange them as shown below.
3. Arrange the recharge on corresponding dates as shown below :

July	Recharge m	C_t values					
		20	19	18	17	16	15
20	0	0.99	-	-	-	-	-
19	0	0.86	0.99	-	-	-	-
18	0.005	0.68	0.86	0.99	-	-	-
17	0.010	0.53	0.68	0.86	0.99	-	-
16	0.020	0.41	0.53	0.68	0.86	0.99	-
15	0.005	0.32	0.41	0.53	0.68	0.86	0.99

4. Multiply the C_t values of each column with the corresponding recharges, adding the results, and dividing the sum by f, which gives the height of water table at the date for which column is taken, as shown below :

July	height of water table, m	
20	20(0.68X0.005 + 0.53X0.01 + 0.41X0.02 + 0.32X0.005)	= 0.37 m
19	20(0.86X0.005 + 0.68X0.01 + 0.53X0.02 + 0.41X0.005)	= 0.48 m
18	20(0.99X0.005 + 0.86X0.01 + 0.68X0.02 + 0.53X0.005)	= 0.60 m
17	20(0.99X0.01 + 0.86X0.02 + 0.68X0.005)	= 0.61 m
16	20(0.99X0.02 + 0.86X0.005)	= 0.48 m
15	20(0.99X0.005)	= 0.10 m

5. Similar method is used for discharge calculation as shown below :

July	Recharge	G_t values					
	m	20	19	18	17	16	15
20	0	0.359	-	-	-	-	-
19	0	0.148	0.359	-	-	-	-
18	0.005	0.110	0.148	0.359	-	-	-
17	0.010	0.085	0.110	0.148	0.359	-	-
16	0.020	0.066	0.085	0.110	0.148	0.359	-
15	0.005	0.051	0.066	0.085	0.110	0.148	0.359

6. Multiplying the G_t values of each column with corresponding recharges and adding the results, yields the total discharge rate at the appropriate date as shown below.

July	drain discharge, m/day	
20	20(0.110X0.005 + 0.085X0.01 + 0.066X0.02 + 0.51X0.005)	= 0.0030 m/day
19	20(0.148X0.005 + 0.110X0.01 + 0.085X0.02 + 0.066X0.005)	= 0.0039 m/day
18	20(0.359X0.005 + 0.148X0.01 + 0.110X0.02 + 0.085X0.005)	= 0.0059 m/day
17	20(0.359X0.01 + 0.148X0.02 + 0.110X0.005)	= 0.0059 m/day
16	20(0.359X0.02 + 0.148X0.005)	= 0.0079 m/day
15	20(0.399X0.005)	= 0.0018 m/day

4.6 van Schilfgaarde Model

van Schilfgaarde (1965) independently but similar to de Zeeuw- Hellinga model, derived another model for simulation of water table behavior in response to rainfall or irrigation input. He considered the **Kirkham (1958)** steady state drain Equation (2.191) in which flow above drain level is neglected to show the non-steady response to periodic rainfall or irrigation. In this equation the drain discharge q is linearly related to the mid point water table.

$$q = \frac{K}{L.F_K} h \tag{4.54}$$

The variation of drain discharge with time is thus also linearly related to the variation in time of the water table head.

$$\frac{dq}{dt} = \frac{K}{LF_K}\frac{dh}{dt} \tag{4.55}$$

If the ground water body is recharged by infiltrated rain or irrigation or another source (EXS) and depleted by drain discharge (q), then water table will rise when (EXS-q)>0 and fall when (EXS-q)<0.

Following the analogy of de Zeeuw and Hellinga (Equation 4.3) the water table fluctuation can be described as.

$$\frac{dh}{dt} = \frac{EXS - q}{Cf} \tag{4.56}$$

where,

C = shape factor or correction factor required to account for steady state assumption. **Bouwer and Schilfgaarde (1963)** suggested that,

$0.02 < h_o/L < 0.08, \quad C = 0.08$

$h_o/L > 0.15, \quad C = 1.00$

Combining Equation (4.55) and Equation (4.56) will yield,

$$\frac{dq}{dt} = \frac{K}{LF_k}\frac{EXS - q}{Cf} = \frac{EXS - q}{A} \tag{4.57}$$

where,

$$A = f.F_k.L.C/K$$

So the change in drain discharge (dq/dt) is proportional to the excess recharge (EXS-q), the constant of proportionality being 1/A. Integrating Equation (4.57) between limit, $t = t$, $q = q_t$, and $t = t-1$, $q = q_{t-1}$ will give,

$$\int_{q_{t-1}}^{q_t} \frac{dq}{EXS - q} = \int_{t-1}^{t} \frac{1}{A} dt$$

(4.58)

$$\frac{(EXS - q)_t}{(EXS - q)_{t-1}} = e^{-\frac{1}{A}\Delta t}$$

Which may be elaborated as,

$$q_t = q_{t-1} e^{-\frac{1}{A}\Delta t} + EXS_{\Delta t}(1 - e^{-\frac{1}{A}\Delta t})$$

(4.59)

In which $ExS_{\Delta t}$ is the mean value of recharge during time interval t-1 to t, and assumed constant. Using the linear relationship between q and h of Equation (4.54), q = (1/A).C. f.. h, it follows :

$$h_t = h_{t-1} e^{-\frac{1}{A}\Delta t} + \frac{EXS_{\Delta t}}{Cf}(1 - e^{-\frac{1}{A}\Delta t})$$

If time interval Δt = 1 day and C =1 above equation reduces to:

$$h_t = h_{t-1} e^{-\frac{1}{A}} + \frac{A}{f} EXS_t (1 - e^{-\frac{1}{A}})$$

And finally equation becomes,

$$h_t = [h_{t-1} e^{-\frac{1}{A}} + \frac{A}{f}(e^{\frac{1}{A}} - 1) EXS_t] e^{-\frac{1}{A}}$$

(4.60)

This is van Schilfgaarde model which can be used for predicting water table behavior in tile drained plots.

4.7 Input to the Drainage System

Amount of water percolating through soil profile is the recharge or input to the drainage system, (EXS_t).

Wiser and Schilfgaarde (1964) gave a water balance model for predicting day to day water balance parameters for given climatological data, soil properties and design parameters.

$$SMC_t = SMC_{t-1} + P_t + I_t - ET_t$$

$$SMC_t \leq MSC \text{ (for all t)} \tag{4.61}$$

where,

SMC_t = soil moisture content at the end of tth day, [L]

SMC_{t-1} = soil moisture content at the end of $(t-1)^{th}$ day, [L]

P_t = effective precipitation (rainfall-surface runoff), [L]

I_t = irrigation amount during day, t, [L]

ET_t = evapotranspiration amount during day, t, [L]

MSC = maximum soil moisture storage capacity, [L]

If the sum on right hand side in Equation (4.61) is greater than MSC, the amount of excess (EXS_t) is taken, as percolation downward to water table and SMC_t is set equal to MSC.

For a given geometry of the drainage system the parameters of Equation (4.60) are fixed and the equation can be solved on a daily basis for the value of EXSt obtained from application of the water balance model.

4.8 Application of van Schilfgaarde Model

Example 4.5

Following rainfall and surface runoff data were recorded from a tile-drained plot having 75 m tile drain spacing at Sampla village in Rohtak district of Haryana state.

Day	Rainfall, (mm)	Surface runoff (mm)	Evapotranspiration (mm)
0	0	0	8
1	100	50	6
2	80	40	5
3	0	0	10
4	0	0	9
5	60	20	5
6	110	10	6
7	0	0	8
8	0	0	8
9	0	0	9
10	0	0	8

Following soil hydrological parameters were worked out in an experiment.

Saturated hydraulic conductivity, K = 4.10 m/day

Transmissivity, $K.d_e$ = 4.91 m^2/day

Drainable porosity, f = 0.14

Drain radius, r = 10 cm

Depth of impervious layer below drain axis, D = 1.2 m

Depth of drain axis below ground surface = 1.75 m

Depth of root zone = 1.0 m

Maximum soil moisture storage capacity, MSC = 330 mm

Soil moisture content before the rain = 305 mm

If the initial water table is at drain level, calculate the water table elevation on daily basis using van Schilfgaarde model.

Solution 4.5

1. Using the **Table 2.3** work out the value of F_K, which comes out to be 8.214.

2. Calculate effective precipitation by subtracting surface runoff amount from rainfall amount.

Day	Rainfall mm	Surface runoff mm	Effective rainfall mm
0	0	0	0
1	100	50	50
2	80	40	40
3	0	0	0
4	0	0	0
5	60	20	40
6	110	10	100
7	0	0	0
8	0	0	0
9	0	0	0
10	0	0	0

3. Now using Wiser and Schilfgaarde's water balance equation, work out drainage input EXSt , on daily basis as below:

$$SMC_t = SMC_{t-1} + I_t + P_t - ET_t$$

Since there is no irrigation, therefore above equation may be also written as,

$$SMC_t = SMC_{t-1} + P_t - ET_t$$

For t = 1 day

$SMC_1 = SMCo + P_1 - ET_1$

$SMC_1 = 305 + 50 - 6 = 49$ mm

Since, $SMC_1 > MSC$

$EXS_1 = 349 - 330 = 19$ mm

For t = 2 day

$SMC_2 = SMC_1 + P_2 - ET_2$

$SMC_2 = 330 + 40 - 5 = 365$ mm

Since, $SMC_2 > MSC$

$EXS_2 = 365 - 330 = 35$ mm

For t = 3 day

$SMC_3 = SMC_2 + P_3 - ET_3$

$SMC_3 = 330 + 0 - 10 = 320$ mm

Since, SMC < MSC

$EXS_3 = 0$ mm

For t = 4 day

$SMC_4 = SMC_3 + P_4 - ET_4$

$SMC_4 = 320 + 0 - 9 = 311$ mm

Since, $SMC_4 < MSC$

$EXS_4 = 0$ mm

For t = 5 day

$SMC_5 = SMC_4 + P_5 - ET_5$

$SMC_5 = 311 + 40 - 5 = 346$ mm

Since, $SMC_5 > MSC$

$EXS_5 = 346 - 330 = 16$ mm

For t = 6 day

$SMC_6 = SMC_5 + P_6 - ET_6$

$SMC_6 = 330 + 100 - 6 = 424$ mm

Since, $SMC_2 > MSC$

$EXS_6 = 424 - 330 = 94$ mm

4. With the help of soil hydrological parameters make following calculations:

$A = f.\ F_K.\ C.\ L / K = 0.14 \text{ X } 8.214 \text{ X } 0.8 \text{ X } 75 \div 4.10 = 16.83$

$A/f = 16.83 \div 0.14 = 120.21$

$(e^{1/A} - 1) = 0.06122$

$e^{-(1/A)} = 0.9423$

5. Substitute these values of parameters in Equation (4.60) one will get,

$$h_t = 0.9423\ h_{t-1} + 6.9346\ EXS_t$$

7. Calculate water table elevation using above equation on daily basis as below:

For t = 1 day,

$$h_1 = 0.9423\ h_o + 6.9346\ EXS_1$$

$$h_1 = 0.9423 \times 0 + 6.9346 \times 19 \times 10^{-3} = 0.132 \text{ m}$$

For t = 2 day,

$$h_2 = 0.9423\ h_1 + 6.9346\ EXS_2$$

$$h_2 = 0.9423 \times 0.132 + 6.9346 \times 35 \times 10^{-3} = 0.367 \text{ m}$$

For t = 3 day,

$$h_3 = 0.9423\ h_2 + 6.9346\ EXS_3$$

$$h_3 = 0.9423 \times 0.367 + 6.9346 \times 0 = 0.346 \text{ m}$$

For t = 4 day,

$$h_4 = 0.9423\ h_3 + 6.9346\ EXS_4$$

$$h_4 = 0.9423 \times 0.346 + 6.9346 \times 0 = 0.326 \text{ m}$$

For t = 5 day,

$$h_5 = 0.9423\ h_4 + 6.9346\ EXS_5$$

$$h_5 = 0.9423 \times 0.326 + 6.9346 \times 16 \times 10^{-3} = 0.418 \text{ m}$$

For t = 6 day,

$$h_6 = 0.9423\ h_5 + 6.9346\ EXS_6$$

$$h_6 = 0.9423 \times 0.418 + 6.9346 \times 94 \times 10^{-3} = 1.046 \text{ m}$$

For t = 7 day,

$$h_7 = 0.9423\ h_6 + 6.9346\ EXS_7$$

$$h_7 = 0.9423 \times 1.046 + 6.9346 \times 0 = 0.986\ m$$

For t = 8 day,

$$h_8 = 0.9423\ h_7 + 6.9346\ EXS_8$$

$$h_8 = 0.9423 \times 0.986 + 6.9346 \times 0 = 0.929\ m$$

For t = 9 day,

$$h_9 = 0.9423\ h_8 + 6.9346\ EXS_9$$

$$h_9 = 0.9423 \times 0.929 + 6.9346 \times 0 = 0.875\ m$$

For t = 10 day,

$$h_{10} = 0.9423\ h_9 + 6.9346\ EXS_{10}$$

$$h_{10} = 0.9423 \times 0.875 + 6.9346 \times 0 = 0.825\ m$$

4.9 Transformed Ernst Model

Chhedi Lal et al. (1989) transformed the **Ernst (1956)** steady state drain spacing equation to transient conditions using de Zeeuw- Hellinga approach, ignoring the vertical resistance. The transformed model is quite simple and does not require correction for radial flow in the near vicinity of the drains. Ernst equation can be written as below:

$$h = \left(\frac{L^2}{8\sum (KD)_h} + \frac{1}{\pi K_r} \ln \frac{aD_r}{u}\right) q \tag{4.62}$$

For homogeneous soil [$\Sigma(KD)_h = 8KD$, $K_h = K_r = K$ and $a = 1$] above equation may be also written as under:

$$h = \left(\frac{L^2}{8KD} + \frac{L}{\pi K} \ln \frac{D}{u}\right) q \tag{4.63}$$

Differentiating Equation (4.63) with respect to time will yield.

$$\frac{dh}{dt} = \left(\frac{L^2}{8KD} + \frac{L}{\pi K}\ln\frac{D}{u}\right)\frac{dq}{dt}$$

(4.64)

On the analogy of de Zeeuw and Hellinga Equation (4.64) may be also written as,

$$\frac{dq}{dt} = \frac{1}{\left(\frac{L^2}{8KD} + \frac{L}{\pi K}\ln\frac{D}{u}\right)}\frac{R-q}{Cf}$$

(4.65)

Upon integration under the limit of t = t, $q_t = q_t$ and t = t-1, $q_t = q_{t-1}$, above equation becomes,

$$\frac{(R-q)_t}{(R-q)_{t-1}} = e^{-\lambda\Delta t}$$

(4.66)

where,

$$\lambda = \frac{1}{Cf\left(\frac{L^2}{8KD} + \frac{L}{\pi K}\ln\frac{D}{u}\right)}$$

Which can be elaborated as,

$$q_t = q_{t-1}e^{-\lambda\Delta t} + R_{\Delta t}(1 - e^{-\lambda\Delta t})$$

(4.67)

Using the linear relationship of Equation (4.63) one can derive the following expression for water table elevation:

$$h_t = h_{t-1}e^{-\lambda\Delta t} + \frac{R_{\Delta t}}{\beta}(1-e^{-\lambda\Delta t})$$

(4.68)

where,

$$\beta = C.f.\lambda$$

Above equation may be used for calculation of water table elevations in response to steady recharge like van Schilfgaarde model.

4.10 Application of Transformed Ernst Model

Problem 4.6

Using the given data set of Problem 4.5, calculate daily water table positions using transformed Ernst's model.

Solution 4.6

Given: Saturated hydraulic conductivity, K = 4.10 m/day

Drainable porosity, f = 0.14

Drain radius, r = 0.05 cm,

Depth of impervious layer below drain axis, D = 1.2 m

1. Calculate the value of λ as under,

$$\lambda = \frac{1}{Cf(\frac{L^2}{8KD} + \frac{L}{\pi K}\ln\frac{D}{u})} = \frac{1}{0.8x0.14(\frac{75^2}{8x4.1x1.2} + \frac{7.5}{\pi x4.1}\ln\frac{1.2}{\pi x0.05})} = 0.0576$$

2. Calculate the value as below :

   β=C.f.λ=0.8 x 0.14 x 0.0576 = 6.4620x10^{-3}:

3. Substitute these values in Equation (4.68) and consider one time interval (Δt = 1 day).

$$h_t = h_{t-1}e^{-0.0576x1} + \frac{R_t}{6.462x10^{-3}}(1-e^{-0.0576x1})$$

$$h_t = 0.944h_{t-1} + 8.662\ R_t$$

4. Considering the input to the drainage EXSt as effective recharge R_t in above equation, water table elevation can be worked out in the following way :

   For t = 1 day,

   $$h_1 = 0.944\ h_0 + 8.662\ R_1$$

   $$h_1 = 0.944 \times 0 + 8.662 \times 19 \times 10^{-3} = 0.17\ m$$

   For t = 2 day,

   $$h_2 = 0.944\ h_1 + 8.662\ R_2$$

   $$h_2 = 0.944 \times 0.17 + 8.662 \times 35 \times 10^{-3} = 0.46\ m$$

   For t = 3 day,

   $$h_3 = 0.944\ h_2 + 8.662\ R_3$$

   $$h_3 = 0.944 \times 0.46 + 8.662 \times 0 = 0.43\ m$$

   For t = 4 day,

   $$h_4 = 0.944\ h_3 + 8.662\ R_4$$

   $$h_4 = 0.944 \times 0.43 + 8.662 \times 0 = 0.41\ m$$

   For t = 5 day,

   $$h_5 = 0.944\ h_4 + 8.662\ R_5$$

   $$h_5 = 0.944 \times 0.41 + 8.662 \times 16 \times 10^{-3} = 0.47\ m$$

   For t = 6 day,

   $$h_6 = 0.944\ h_5 + 8.662\ R_6$$

   $$h_6 = 0.944 \times 0.47 + 8.662 \times 94 \times 10^{-3} = 1.29\ m$$

   For t = 7 day,

   $$h_7 = 0.944\ h_6 + 8.662\ R_7$$

   $$h_7 = 0.944 \times 1.29 + 8.662 \times 0 = 1.22\ m$$

   For t = 8 day,

   $$h_8 = 0.944\ h_7 + 8.662\ R_8$$

$h_8 = 0.944 \text{ x } 0.1.22 + 8.662 \text{ x } 0 = 1.15 \text{ m}$

For t = 9 day,

$h_9 = 0.944\, h_8 + 8.662\, R_9$

$h_9 = 0.944 \text{ x } 1.15 + 8.662 \text{ x } 0 = 1.08 \text{ m}$

For t = 10 day,

$h_{10} = 0.944\, h_9 + 8.662\, R_{10}$

$h_{10} = 0.944 \text{ x } 1.08 + 8.662 \text{ x } 0 = 1.02 \text{ m}$

References

Chhedi Lal, Sewa Ram, Chauhan, H.S. and K.V.G.K., Rao.1989. Transformation of Ernst's steady state solution to predict unsteady water table for subsurface drainage. Journal of Indian Water Resources Society, 9(4):30-37.

de Zeeuw, J.W. and F. Hellinga.1958. Neerslag en afvoer. Landbouwkundig Tijdschrift 70:405-421.

de Zeeuw, J.W.1966. Anlyse van het afvoerverloop van bebieden met hoofdzakelijk grondwaterafvoer. Mede. Landbouwhogeschool Wageningen, 66-5. 139 p.

Hellinga, F.1952. Enige aspecten van de inrichting van polders. De Ingeneur, 64:B33-B36.

Ernst, L.F.1956. Calculation of the steady state flow of ground water in vertical cross section.Neth. J. Agric. Science. 4:126-131

Kirkham, D.1958. Seepage of steady rainfall through soil into drains. Trans. Am. Geophys. Union 39:892-908.

Kraijenhoff van de Leur, D.A.1958. A study on non-steady ground water flow with special reference to a reservoir coefficient. De Ingenieur 40:87-94.

Maasland, M.1959. Water table fluctuations induced by intermittent recharges. Journal Geophys. Res. 64:549-559.

van Schilfgaarde, J. 1965. Transient design of drainage system. Proc. Am. Soc. Civil Engng. 91(IR-3):9-22.

Wiser, E.H. and J. van Schilfgaarde.1964. Prediction of irrigation water requirements using water balance. Proc. 6th Cong. Int. de Genie Rurals, 121-129.

Wiser, E.H., Robert, C.W. and D.A.Link. 1974.Optimised design of a subsurface drainage system. Trans. of the ASAE, 17(1):175-178, 182.

Problems

Theoretical

1. Describe the de Zeeuw and Hellinga (1958) approach to transform steady state drain spacing formula for predicting drain discharge and water table height.
2. Write short comments on (1) reservoir coefficient, and (2) reaction factor.
3. Write the water balance modelof Wiser and Schilfgaarde (1964) and discuss about the parameters, briefly.
4. Derive de Zeeuw and Hellinga model for predicting transient drain discharges in response to constant recharge.
5. Transform the de Zeeuw-Hellinga's transient drain discharge equation to predict transient water table height midway between the drains in response to recharge.
6. If recharge becomes zero, what shape the de Zeeuw -Hellinga model takes place.
7. Derive Kraijenhoff van de Leur-Maasland equation for predicting transient water table height midway between the drains in response to raifall/recharge.
8. Transform Ernst steady state drain spacing formula to transient conditions to predict drain discharge in response to recharge using de Zeeuw-Heelinga approach.

Numerical

1. There was a total effective rainfall of 50 mm which percolated to ground water and then there was no rainfall (no percolation) in following days. What would have been the height of midway table between the drain after four days of rainfall, if the initial water level was at drain level, spacing between the drain was 20 m, drainable porosity 0.05, depth to impervious layer 1.2 m and equivalent depth of impervious layer is 1.0 m.

   [Answer: 0.53 m]

2. A drainage system with j = 4 days in a soil with effective porosity f = 0.05 receives the following ground water recharge.

January	10	11	12	13	14	15
Recharge (mm)	5	0	0	10	5	20

   What height of water table will occur on each day if the initial water table height is at drain level.

   [Answer: 0.73 m, 0.35m, 0.45m, 0.56m, 0.44m, 0.40m]

3. Following data were collected from a tile-drained plot:

   Drain spacing = 40 m, depth of drain below ground surface = 1.2 m,

   Transmissivity = 2.5 m^2/day, drainable porosity = 0.05

   Calculate the water table height midway between the drains for 10 days for the following recharge pattern if the initial water table height above drain is 0.20 m.

July	1	2	3	4
Recharge, mm	35	0	0	10

   [Answer: 0.93 m, 0.68m, 0.50m, 0.59m, 0.43m, 0.31m, 0.23m, 0.17m, 0.12m, 0.9m,]

4. Following data were collected from a tile drained plot located at Sampla village in Rohtak district of Haryana.

   a) Transmissivity, Kd = 4.91 m^2/day

b) Drainable porosity, f = 0.14

c) Drain radius, r = 0.10 m

d) Depth if impervious layer below drains axis, D = 1.2 m

e) Depth of drain axis below ground surface = 1.75 m

f) Depth of root zone = 1.0 m

g) Maximum soil moisture storage capacity, MSC = 330 mm

h) Saturated hydraulic conduct ivity, K = 4.1m/day

i) Spacing between the drain, L = 75m

Using van Schilfgaarde model predict water table height for next 10 days, if the initial water level is 0.50 m above drain axis for the following recharge pattern :

Table (day)	Effective rainfall/input to drainage system, EXSt (mm)
1	0
2	40
3	30
4	0
5	10
6	80
7	0

[Answer : 0.47m, 0.72m, 0.89m, 0.84m, 0.86m, 1.36m, 1.28m, 1.21m, 1.14m, 1.08m,]

chapter 5

DRAINAGE OF SLOPING LAND

A large number of drainage theories, which have been developed so far, pertain to equally spaced drains located in a flat land. There are, however, in different parts of the world many areas of sloping lands also that require subsurface drainage. The problem of drainage is generally analysed considering uniform rate of replenishment. However, time variable replenishment rates of the forms of linearly increasing, exponentially declining and a product of linear and exponential form are also common in nature. In the present chapter, theoretical analyses carried out by different workers are discussed.

Sewa Ram and Chauhan (1985) solved the fundamental Boussinesq equation, which describes the ground water flow for sloping lands, for constant rate and time variable replenishment rate and approximate boundary conditions. The system has been shown in **Figure 5.1**.

5.1 Assumptions Involved

Following assumptions were made to simplify the water flow problem in sloping land.

1. The phreatic aquifer lies over a moderately sloping impermeable bed.
2. Soil is homogeneous, isotropic and semi-infinite.
3. The rate of replenishment is less than the rate of hydraulic conductivity of soil.
4. The deep percolation moves vertically downward until it joins the main ground water body.
5. Dupuit-Forchheimer assumptions hold good.
6. Soil is drained by equally spaced parallel drains.
7. The drains cut through the depth of phreatic aquifer up to the impermeable layer.

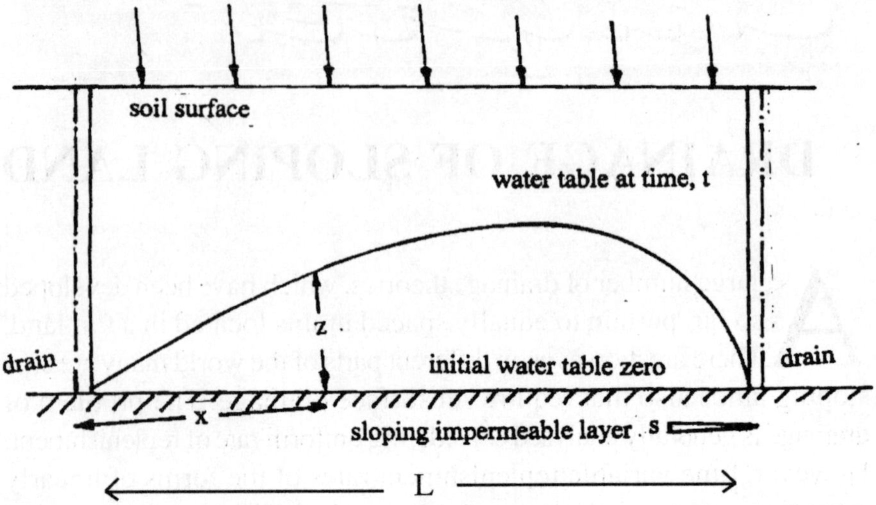

Figure 5.1. Definition sketch for ground water flow over a sloping impermeable layer receiving replenishment.

5.2 Governing Differential Equation

The flow system is defined by the following equation, originally derived by Boussinesq (1904) for ground water flow in a gently sloping aquifer above an impermeable and later modified by Werner (1953, 1957) to account for the replenishment rate factor also.

$$K\left[z\frac{\partial^2 z}{\partial x^2}+\left(\frac{\partial z}{\partial x}\right)^2\right]-K.S_l.\frac{\partial z}{\partial x}=f\frac{\partial z}{\partial t}-p \qquad 5.1$$

where,

f = drainable porosity of aquifer, $[L^3/L^3]$

K= saturated hydraulic conductivity of aquifer, [L/T]

x = horizontal distance away from the drain, [L]

z = height of the phreatic surface above impervious layer, [L]

S_l= slope of impermeable layer, [L/L]

p = recharge or replenishment rate, [L/T]

Equation 5.1 with a time variable replenishment rate may be written as below :

$$K\left[z\frac{\partial^2 z}{\partial x^2}+\left(\frac{\partial z}{\partial x}\right)^2\right]-K.S_l.\frac{\partial z}{\partial x}+p(t)=f\frac{\partial z}{\partial t}$$

(5.2)

Equation (5.2) is a non-linear second order partial differential equation and is generally solved by the simplifications of neglecting the term $(\partial z./\partial x)^2$.

5.3 General Initial and Boundary Conditions

The water table is considered at the drain level before the replenishment takes place, which is the most practical initial condition. The boundary conditions are considered zero at all times. The effect of seepage surface is not considered pronounced and hence ignored. The initial and boundary conditions may be written as follows :

Initial condition: $z(x, o) = 0$

Boundary condition: $z(0, t) = 0 = z(L, t)$

$p_{(t)} = 0$ at $t \leq 0$, $p = p_{(t)}$ at $t \geq 0$ (5.3)

5.4 Recharge/Replenishment Rate

The replenishment is the rate of surface applied recharge per unit area of drained land. The following four types of arbitrary selected replenishment rates have been considered under present analysis :

1. Constant rate, $\mathbf{p_{(t)} = p}$
2. Linearly increasing rate, $\mathbf{p_{(t)} = mt}$
3. Exponentially declining rate, $\mathbf{p_{(t)} = pe^{-rt}}$
4. Product of linearly increasing and exponentially declining rate, $\mathbf{p_{(t)} = (p\text{-}mt)e^{-rt}}$

5.5 Transformation of Governing Equation

Werner (1953, 1957) used a transformation which describes the term $(\partial z/\partial x)^2$ in differential equation. The transformation is,

$$z^2 = y$$

Which transform the Equation (5.2) into the following form:

$$\frac{\partial^2 y}{\partial x^2} - \frac{S_l}{D}\frac{\partial y}{\partial x} = \frac{f}{D}\frac{\partial y}{\partial t} - \frac{2p_{(t)}}{K} \tag{5.4}$$

Where D is weighted mean depth of the saturated aquifer (Dumm, 1954), replaced for the variable z to linearize the transformed equation. The assumption of the weighted mean depth, D for the variable z has been used by a number of investigators and leads to satisfactory results if the drains completely penetrate the flow medium. If the flow converges near partially penetrating drains needs to be accounted for, D may be evaluated by a method described by Maasland (1956).

Equation (5.4) along with the initial and boundary conditions (5.3) after transformation, may be written as,

$$\frac{\partial^2 y}{\partial x^2} - 2a\frac{\partial y}{\partial x} = \frac{1}{c}\frac{\partial y}{\partial t} - \frac{2p_{(t)}}{K} \tag{5.5}$$

where,

$$a = S_l/(2KD)$$
$$c = KD/f \tag{5.6}$$

Transformed initial and boundary conditions are written as below:

$$y\,(x, o) = 0$$
$$y\,(0, t) = 0 = y\,(L, t)$$
$$p_{(t)} = 0 \quad \text{at} \quad t \le 0, \text{ and}$$
$$p = p_{(t)} \quad \text{at} \quad t \ge 0 \tag{5.7}$$

5.6 Analytical Solutions

Equation (5.5) with initial and boundary conditions (5.7) has been solved for different types of replenishment rates outlined earlier, one by one.

5.6.1 Steady state flow in a sloping aquifer with constant replenishment

5.6.1.1 Governing equation

The solution for steady state flow in a sloping aquifer receiving constant replenishment rate, p, can be obtained from Equation (5.4) which reduces to following form after transformation for steady state conditions:

$$\frac{\partial^2 y}{\partial x^2} - \frac{S_l}{D}\frac{\partial y}{\partial x} + \frac{2p}{K} = 0 \tag{5.8}$$

5.6.1.2 Solution

Boundary conditions for constant replenishment rate are written below:

y = 0, at x = 0

y = 0 at x = L (5.9)

Let s = S_l/D (5.10)

w = 2p/K (5.11)

Equation (5.8) becomes as follows with changed conditions:

$$\frac{\partial^2 y}{\partial x^2} - S_l\frac{\partial y}{\partial x} + w = 0 \tag{5.12}$$

The general solution of Equation (5.12) may be obtained as,

$$y = (C_! + C_2 e^{sx}) + \frac{wx}{s} \tag{5.13}$$

Now using the boundary conditions (5.9) C_1 and C_2 can be obtained as under.

$$y = 0 \text{ at } x = 0,$$

$$0 = (C_1 + C_2 e^{-sx0}) + \frac{wx0}{s}$$

$$0 = C_1 + C_2$$

$$C_1 = -C_2$$

(5.14)

$y = 0$, at $x = L$,

$$0 = C_1 + C_2 e^{sL} + \frac{wL}{s}$$

$$= -C_2 + C_2 e^{sL} + \frac{wL}{s}$$

$$C_2 - C_2 e^{sL} = \frac{wL}{s}$$

(5.15)

Substituting these values of C_1 and C_2 in Equation (5.13) one will get.

$$y = \left(-\frac{wL}{s(1-e^{sL})} + \frac{wL}{s(1-e^{sL})} e^{sx} \right) + \frac{wx}{s}$$

$$= \frac{w}{s}\left[x - \frac{L(e^{sx}-1)}{(e^{sL}-1)} \right]$$

(5.16)

Since $y = z^2$, therefore,

$$z^2 = \frac{w}{s}\left[x - \frac{L(e^{sx}-1)}{(e^{sL}-1)} \right]$$

(5.17)

Equation (5.17) gives the steady state profile of water table in a sloping aquifer receiving constant replenishment rate.

As a special case, for zero slope, the solution for steady state of water table can be obtained from Equation (5.17). By expanding the exponential terms e^{sx} and e^{sL} appearing in Equation (5.17) in series, neglecting higher order, i.e. more than square, terms and simplifying one will obtain.

$$z^2 = \frac{w}{2} x \frac{(L-x)}{(1+\frac{sL}{2})} \tag{5.18}$$

where s = 0, for zero slope, thus,

$$z^2 = \frac{w}{2} x(L-x) = \frac{p}{K} x(L-x) \tag{5.19}$$

Equation (5.19) describes the steady state flow for horizontal aquifers receiving constant replenishment rate. For midpoint Equation (5.19) reduces to,

$$z^2 = \frac{pL^2}{4K} \tag{5.20}$$

which shows the water table to be of elliptic shape which is the well known steady state ellipse equation, first derived by Hooghoudt (1940).

5.6.2. Constant replenishment rate, p(t) = p

5.6.2.1 Governing equation

The problem is defined by Equation (5.5) and initial and boundary conditions (5.7) with p(t) = p. The governing differential equation along with initial and boundary conditions is thus given by,

$$\frac{\partial^2 y}{\partial x^2} - 2a\frac{\partial y}{\partial x} = \frac{1}{c}\frac{\partial y}{\partial t} - \frac{2p}{K} \tag{5.21}$$

5.6.2.2 Initial and boundary conditions

$$y(x,0) = 0$$

$$y(0,t) = 0 = y(L,t)$$

$$p=0 \text{ at } \quad t \leq 0; \quad p = p \text{ at } t \geq 0$$

5.22)

5.6.2.3 Transformation of governing equation

A transformation was devised as written below.

$$y = Ve^{ax-ca^2t} + \frac{2cpt}{K} \tag{5.23}$$

Substituting above equation in Equation (5.21)

$$\frac{dy}{dt} = \frac{\partial V}{\partial t}.e^{ax-ca^2t} + V(-ca^2)e^{ax-ca^2t}\frac{dt}{dt}$$

$$\frac{dy}{dt} = \frac{\partial V}{\partial t}e^{ax-ca^2t} - ca^2Ve^{ax-ca^2t} \tag{5.24}$$

$$\frac{\partial y}{\partial x} = \frac{\partial V}{\partial x}.e^{ax-ca^2t} + V.a.e^{ax-ca^2t}\frac{dx}{dx}$$

$$\frac{\partial y}{\partial x} = \frac{\partial V}{\partial x}e^{ax-ca^2t} + a.Ve^{ax-ca^2t} \tag{5.25}$$

$$\frac{\partial y}{\partial x} = (\frac{\partial^2 V}{\partial x^2}e^{ax-ca^2t} + \frac{\partial V}{\partial x}a.e^{ax-ca^2t} + a\frac{\partial V}{\partial x}e^{ax-ca^2t} + a^2Ve^{ax-ca^2t}$$

$$\frac{\partial^2 y}{\partial x^2} = \frac{\partial^2 V}{\partial x^2}e^{ax-ca^2t} - a^2Ve^{ax-ca^2t} + 2\frac{\partial V}{\partial x}ae^{ax-ca^2t} \tag{5.26}$$

Substituting these values in Equation (5.21)

$$\frac{\partial^2 V}{\partial x^2} e^{ax-ca^2t} + a^2.V.e^{ax-ca^2t} + 2a\frac{\partial V}{\partial x} e^{ax-ca^2t} - \frac{1}{c}\frac{\partial V}{\partial t} e^{ax-ca^2t} + ca^2\frac{V}{c} e^{ax-ca^2t} = 0$$

We finally get,

$$\frac{\partial^2 V}{\partial x^2} - \frac{1}{c}\frac{\partial V}{\partial t} = 0 \tag{5.27}$$

5.6.2.4 Transformation of initial and boundary conditions

The initial and boundary conditions after transformation reduces to,

$$V(x,0) = 0$$

$$V(0,t) = -\frac{2cpt}{K} e^{ca^2t}$$

$$V(L,t) = -\frac{2cpt}{K} e^{-aL} + ca^2t$$

$$\mathrm{p} = 0, \quad \text{at } t<0, \qquad \mathrm{p} = \mathrm{p}, \text{ at } t \geq 0 \tag{5.28}$$

5.6.2.5 Solution

The solution of Equation (5.27) with the general initial and boundary conditions,

$$\begin{aligned} V(x,0) &= 0 \\ V(o,t) &= \phi_1(\lambda) \\ V(L,t) &= \phi_2(\lambda) \end{aligned} \tag{5.29}$$

is given in Carlslaw and Jaeger (1959) as below:

$$V = \frac{2c\pi}{L^2} \sum_{n=1,2,3,..}^{\infty} (ne^{-c(\frac{n\pi}{L})^2 t} . \sin\frac{n\pi}{L} x)[\int_0^t e^{c(\frac{n\pi}{L})^2 \lambda} \{\phi_1(\lambda) - (-1)^n \phi_2(\lambda)\} d\lambda \tag{5.30}$$

The solution of Equation (5.27) for the initial and boundary conditions (5.28) may be obtained from Equation (5.30) as below:

$$V = \frac{2\pi c}{L^2} \sum_{n=1,2,3,..}^{\infty} \left((ne^{-c(\frac{n\pi}{L})^2 t}).\sin\frac{n\pi}{L}x \right) \left[\int_0^t e^{c(\frac{n\pi}{L})^2 \lambda} \left\{ \frac{-2cp}{K} e^{ca^2\lambda} + (-1)^n \frac{2cp\lambda}{K} e^{-aL+ca^2\lambda} \right\} d\lambda \right] \tag{5.31}$$

Carrying out integration and simplifying one will get,

$$V = \frac{4\pi c^2 p}{KL^2} \sum_{n=1,2,3,..}^{\infty} \left((ne^{-c(\frac{n\pi}{L})^2 t}).\sin\frac{n\pi}{L}x \right) . [1-(-1)^n e^{aL}] \left[e^{cBt} \left\{ \frac{1}{(cB)^2} - \frac{t}{cB} \right\} - \frac{1}{(cB)^2} \right] \tag{5.32}$$

where

$$B = a^2 + (n\pi/L)^2 \tag{5.33}$$

Solution of Equation (5.21) may be obtained by substituting the inverse of the transform (5.23) as below:

$$y(x,t) = \frac{2cp}{K}\left[t + \frac{2c\pi}{L^2} \sum_{n=1,2,3,..}^{\infty} (ne^{ax}.\sin\frac{n\pi}{L}x).\{1-(-1)^n e^{-aL}\} \left\{ \frac{1}{(cB)^2}(1-e^{-cBt}) - \frac{t}{cB} \right\} \right] \tag{5.34}$$

The solution of Equation (5.3) may be obtained by substituting the inverse of the transformation (z^2=Y) as below:

$$z^2(x,t) = \frac{2cp}{K}\left[t + \frac{2c\pi}{L^2} \sum_{n=1,2,3,..}^{\infty} (ne^{ax}.\sin\frac{n\pi}{L}x).\{1-(-1)^n e^{-aL}\} \left\{ \frac{1}{(cB)^2}(1-e^{-cBt}) - \frac{t}{cB} \right\} \right] \tag{5.35}$$

The Equation (5.35) thus, gives the unsteady state water table profile for sloping lands receiving constant replenishment rate for zero initial and boundary conditions. As a special case for flat land (S_1=0) Equation (5.35) reduces to:

$$z^2(x,t) = \frac{8pL^2}{K\pi^3} \sum_{n=1,3,5,..}^{\infty} n^{-3}.\sin\frac{n\pi}{L}x(1-e^{-c(\frac{n\pi}{L})^2 t}) + \frac{2cpt}{K}(1 - \frac{4}{\pi} \sum_{n=1,3,5,..}^{\infty} n^{-1}\sin\frac{n\pi}{L}x) \tag{5.36}$$

Substituting the identity,

$$\frac{4}{\pi}\sum_{n=1,3,5,..}^{\infty} n^{-1}\sin\frac{n\pi}{L}x \approx 1 \tag{5.37}$$

in the Equation (5.36) one will get,

$$z^2(x,t) = \frac{8pL^2}{K\pi^3}\sum_{n=1,3,5,..}^{\infty} n^{-3}\sin\frac{n\pi}{L}x(1-e^{-c(\frac{n\pi}{L})^2 t}) \tag{5.38}$$

The above solution given by Equation (5.38) can also be obtained from a similar boundary value problem for non-sloping land solved by Maasland (1959).

5.6.3 Linearly increasing replenishment rate, p(t)=mt

5.6.3.1 Governing equation

The differential Equation (5.3) and the initial and boundary conditions (5.7), for the linearly increasing replenishment rate may be written as follows:

$$\frac{\partial^2 y}{\partial x^2} - 2a\frac{\partial y}{\partial x} = \frac{1}{c}\frac{\partial y}{\partial t} - \frac{2mt}{K} \tag{5.39}$$

5.6.3.2 Initial and boundary conditions

Following initial and boundary conditions were used for deriving a solution:

$$\left.\begin{array}{l} y(x,0) = 0 \\ y(0,t) = y(L,t) = 0 \\ p(t) = 0, \quad \text{at } t<0, \qquad p(t) = m.t, \quad \text{at } t\geq 0 \end{array}\right| \tag{5.40}$$

5.6.3.3 Transformation of governing equation

A transformation was devised to transform the governing equation into linear form as below:

$$y = Ve^{ax-ca^2t} + \frac{mc}{K}t^2 \tag{5.41}$$

which transformed Equation (5.39) into the standard heat flow equation as below:

$$\frac{\partial^2 V}{\partial x^2} = \frac{1}{K}\frac{\partial V}{\partial t}$$

5.6.3.4 Transformation of initial and boundary conditions

Initial and boundary conditions using Equation (5.41) as below:

$$\left.\begin{aligned} &V(x,o) = 0 \\ &V(0,t) = -\frac{mc}{K}t^2 e^{ca^2t} \\ &V(L,t) = -\frac{mc}{K}t^2 e^{-aL+ca^2t} \end{aligned}\right\} \tag{5.42}$$

5.6.3.5 Solution

The solution of Equation (5.27) for the initial and boundary conditions (5.42) can be obtained on similar lines as for the constant replenishment case from Equation (5.30) as below.

$$V = \frac{2c\pi}{L^2}\sum_{n=1,2,3,..}^{\infty} n.e^{-c(\frac{n\pi}{L})^2 t}.\sin\frac{n\pi}{L}x\left[\int_0^t e^{c(\frac{n\pi}{L})^2\lambda}\left\{-\frac{mc}{K}\lambda^2 e^{ca^2\lambda} - (-1)^n(-)\frac{mc}{K}\lambda^2 e^{-aL+ca^2\lambda}\right\}d\lambda\right] \tag{5.43}$$

Carrying out the integration, simplifying and rearranging, one will get.

$$V = \frac{2\pi c^2 m}{KL^2}\sum_{n=1,2,3,..}^{\infty}\left((ne^{-c(\frac{n\pi}{L})^2 t}).\sin\frac{n\pi}{L}x\right).[1-(-1)^n e^{-aL}\left[\frac{2}{(cB)^2} - e^{cBt}\left\{\frac{t^2}{cB}\frac{2t}{(cB)^2} + \frac{2}{(cB)^3}\right\}\right]] \tag{5.44}$$

Substituting the inverse of the transformation (5.41) and the transformation $y=z^2$, one will obtain,

$$z^2(x,t)=\frac{mc}{K}[t^2+\frac{2\pi c}{L^2}\sum_0^{\infty}(neax\sin\frac{n\pi}{L}x\{1-(-1)^n e^{-aL}\}.$$

$$\{\frac{2t}{(cB)^2}-\frac{t^2}{cB}-\frac{2}{(cB)^3}(1-e^{-cBt})\}]$$

(5.45)

As a special case, for zero slope the solution can be obtained from Equation (5.45) by taking $S_1=0$ and simplifying as below:

$$z^2(x,t)=\frac{8L^2m}{K\pi^3}\sum_{n=1,3,5,..}^{\infty}(n^{-3}\sin\frac{n\pi}{L}x)\left[t-\frac{1}{c(\frac{n\pi}{L})^2}\left\{1-e^{-c(\frac{n\pi}{L})^2 t}\right\}\right]$$

$$+\frac{mc}{K}t^2(1-\frac{4}{\pi}\sum_{n=1,3,5,..}^{\infty}n^{-1}\sin\frac{n\pi}{L}x)$$

(5.46)

Using the identity y Equation (5.37) Equation (5.46) reduces to,

$$z^2(x,t)=\frac{8L^2m}{K\pi^3}\sum_{n=1,3,5,..}^{\infty}(n^{-3}\sin\frac{n\pi}{L}x)\left[t-\frac{1}{c(\frac{n\pi}{L})^2}(1-e^{-c(\frac{n\pi}{L})^2 t})\right]$$

(5.47)

Equation (5.47) thus gives the solution for unsteady state water table profile for horizontal phreatic aquifer receiving linearly increasing replenishment.

5.6.4 Exponentially declining replenishment, $p(t)=pe^{-rt}$

5.6.4.1 Governing equation

The differential Equation (5.5) for the exponentially declining replenishment rate may be written as:

$$\frac{\partial^2 y}{\partial x^2} - 2a\frac{\partial y}{\partial x} = \frac{1}{c}\frac{\partial y}{\partial t} - \frac{2pe^{-rt}}{K} \tag{5.48}$$

5.6.4.2 Initial and boundary conditions

$$\left.\begin{aligned} & y(x,0)=0 \\ & y(0,t)=0=y(L,t) \\ & p(t)=0 \quad \text{at } t<0 \quad \text{and} \quad p(t)=pe^{-rt} \text{ at } t\geq 0 \end{aligned}\right\} \tag{5.49}$$

5.6.4.3 Transformation of governing equation

A transformation was devised to transform the Equation (5.49) as below:

$$y = Ve^{ax-ca^2t} - \frac{2cp}{Kr}e^{-rt} \tag{5.50}$$

Which transforms the Equation (5.49) into the standard heat flow equation as below:

$$\frac{\partial^2 V}{\partial x^2} = \frac{1}{c}\frac{\partial V}{\partial t}$$

5.6.4.4 Transformation of initial and boundary conditions

The initial and boundary conditions (5.49) can be also transformed using Equation (5.50).

$$\begin{aligned} V(x,0) &= 0 \\ V(0,t) &= \frac{2cp}{Kr}e^{(ca^2-r)t} \\ V(L,t) &= \frac{2cp}{Kr}e^{-aL+(ca^2-r)t} \end{aligned} \tag{5.51}$$

5.6.4.5 Solution

The solution of Equation (5.27) with initial and boundary conditions

$$\left.\begin{array}{l} V(x,0) = f(x) \\ V(0,t) = \phi_1(\lambda) \\ V(L,t) = \phi_2(\lambda) \end{array}\right\} \tag{5.52}$$

is given by Carlslaw and Jaeger (1959) as below:

$$V = \frac{2}{L}\sum_{n=1,2,3,..}^{\infty} e^{-c(\frac{n\pi}{L})^2\lambda} \sin\frac{n\pi}{L}x$$

$$\left[\int_0^t f(x)\sin\frac{n\pi}{L}xdx + \frac{n\pi c}{L}\int_0^t ec(\frac{n\pi}{L})^2\lambda\{\phi_1(\lambda)-(-1)^n\phi_2(\lambda)\}d\lambda\right] \tag{5.53}$$

The solution of Equation (5.27) for the initial and boundary conditions (5.51) can be obtained from Equation (5.53) as below:

$$V = \frac{2}{L}\sum_{n=1,2,3,..}^{\infty} e^{-c(\frac{n\pi}{L})^2 t}.\sin\frac{n\pi}{L}x\left[\int_o^L \frac{2cp}{K.r}e^{-ax}.\sin\frac{n\pi}{L}x + \right.$$

$$\left.\frac{n\pi c}{L}\int_0^t e^{c(\frac{n\pi}{L})^2 t}.\frac{2cp}{K.r}.e^{(ca^2-r)\lambda}\{1-(-1)^n e^{-aL}\}d\lambda\right] \tag{2.54}$$

Carrying out integration and simplification one may get,

$$V = \frac{4cp\pi}{KrL^2}\sum_{n=1,2,3,..}^{\infty}\frac{(n\sin\frac{n\pi}{L}x)}{(cB-r)}[1-(-1)^n e^{-aL}][ce^{(ca^2-r)t} - \frac{r}{B}e^{-c(\frac{n\pi}{L})^2 t}] \tag{5.55}$$

Substituting the inverse of the transformation (5.50) one will get the solution of boundary value problem given by Equation (5.48) with initial and boundary conditions (5.49).

$$z^2(x,t) = \frac{2cp}{K.r}\left[-e^{-rt} + \frac{2}{L^2}\sum_{n=1,2,3,..}^{\infty}\{(ne^{ax}.\sin\frac{n\pi}{L}x)/(cB-r)\}\{1-(-1)^n e^{-aL}\}\left\{ce^{-rt} - \frac{r}{B}e^{-cBt}\right\}\right] \tag{5.56}$$

Equation (5.56) is the solution for unsteady state water table profile in a sloping phreatic aquifer receiving exponentially declining replenishment rate. As a special case for zero slope Equation (5.56) reduces as below:

$$z^2(x,t) = \frac{2cp}{K.r}\left[-e^{-rt} + 4\pi \sum_{n=1,3,5,..}^{\infty} \frac{n.\sin\frac{n\pi}{L}x}{(cn^2\pi^2 - rL^2)}\left\{ce^{-rt} - \frac{rL^2}{n^2\pi^2}.e^{-c(\frac{n\pi}{L})^2 t}\right\}\right] \tag{5.57}$$

5.6.5 Product of linearly increasing and exponentially declining replenishment rate, p(t)=(p-mt)e^{-rt}

5.6.5.1 Governing equation

The boundary value problem for the above case may be formulated using Equation (5.5) as given below:

$$\frac{\partial^2 y}{\partial x^2} - 2a\frac{\partial y}{\partial x} = \frac{1}{c}\frac{\partial y}{\partial t} - \frac{2(p-mt)e^{-rt}}{c} \tag{5.58}$$

5.6.5.2 Initial and boundary conditions

Following initial and boundary conditions have to be imposed on governing equation:

$$\left.\begin{array}{l} y(x,0) = 0 \\ y(0,t) = 0 = y(L,t) \\ p(t) = 0 \quad \text{at } t<0 \quad \text{and} \quad p(t) = (p-mt)e^{-rt} \text{ at } t\geq 0 \end{array}\right\} \tag{5.59}$$

5.6.5.3 Transformation of governing equation

A transformation was devised to transform the Equation (5.58) as below:

$$y = Ve^{ax-ca^2t} - \frac{2cp}{Kr}(\frac{m}{r} - p + mt)e^{-rt} \tag{5.60}$$

Which transforms the Equation (5.58) into the standard heat flow equation as below:

$$\frac{\partial^2 V}{\partial x^2} = \frac{1}{c}\frac{\partial V}{\partial t}$$

5.6.5.4 Transformation of initial and boundary conditions

The initial and boundary conditions (5.59) can be also transformed using Equation (5.60).

$$V(x,0) = \frac{-2c}{K.r}(\frac{m}{r} - p)e^{-ax}$$

$$V(0,t) = -\frac{2cp}{Kr}(\frac{m}{r} - p + mt)e^{(ca^2-r)t}$$

$$V(L,t) = -\frac{2c}{Kr}(\frac{m}{r} - p + mt)e^{-aL+(ca^2-r)t} \tag{5.61}$$

5.6.5.5 Solution

The solution of Equation (5.27) with initial and boundary conditions (5.61) can be obtained on similar lines as for the exponentially declining replenishment case from Equation (5.53) as below:

$$V = \frac{2}{L}\sum_{n=1,2,3,..}^{\infty} e^{\frac{-cn^2\pi^2t}{L^2}}.\sin\frac{n\pi}{L}x[\int_0^L(-)\frac{2c}{Kr}(\frac{m}{r} - p)e^{-ax}.\sin\frac{n\pi}{L}xdx + \frac{n\pi c}{L}.$$

$$\int_0^t ec(\frac{n^2\pi^2\lambda}{L^2})\{(-)\frac{2c}{Kr}(\frac{m}{r} - p + mt)e^{(ca^2-r)\lambda}(1-(-1)^n e^{-aL})\}d\lambda] \tag{5.62}$$

Carrying out integration and simplifying one will get

$$V=\frac{2\pi}{L^2}\sum_{n=1,2,3,..}^{\infty} ne^{-c(\frac{n\pi}{L})^2 t}.\sin\frac{n\pi}{L}x\left[1-(-1)^n e^{aL}\right]\frac{2c}{Kr}[\frac{c}{cB-r}(\frac{m}{cB-r}-mt-G).e^{(cB-r)t}$$

$$\left.\frac{c}{cB-r}(\frac{m}{cB}-mt-G)-\frac{G}{B}\right]$$

(2.63)

$$G=\frac{m}{r}-p$$

(5.64)

Substituting the inverse of the transformation (5.60) one will get the solution of boundary value problem given by Equation (5.63).

Solution of (5.58) with initial and boundary conditions (5.59) in terms of y(x,t) from which the solution of Equation (5.58) can be obtained by putting inverse of transformation $y=z^2$ as below:

$$z^2(x,t)=\frac{2c}{Kr}[(G+mt)e^{-rt}+\frac{2\pi}{L^2}\sum_{n=1,2,3,..}^{\infty}(ne^{ax}\sin\frac{n\pi}{L}x)\{1-(-1)^n e^{-aL}\}$$

$$\{\frac{c}{cB-r}(\frac{m}{cB-r}-mt-G)e^{-rt}-\frac{ce^{-KBt}}{cB-r}(\frac{m}{cB-r}-G)-\frac{Ge^{-cBt}}{B}\}]$$

(5.65)

As a special case for zero slope Equation (5.65) reduces to,

$$z^2(x,t)=\frac{2c}{Kr}[(G+mt)e^{rt}+4\pi\sum_{n=1,3,5,..}^{\infty}(n\sin\frac{n\pi}{L}x)\{\frac{c}{cn^2\pi^2-rL^2}(\frac{mL^2}{cn^2\pi^2-rL^2}$$

$$-mt-G)e^{-rt}-\frac{ce^{-c(\frac{n\pi}{L})^2 t}}{cn^2\pi^2-rL^2}(\frac{mL^2}{cn^2\pi^2-rL^2}-G)-\frac{Ge^{-c(\frac{n\pi}{L})^2 t}}{n^2\pi^2}\}]$$

(5.66)

Sewa Ram and Chauhan (1984) experimentally verified the all derived analytical solutions on a Hele-Shaw model for 0, 2.6, 5.2, and 7.8% slopes. They calculated that:

1. The analytical solutions resulted in more error near the drains than the mid point.
2. The error was more in the initial period of buildup of water table.
3. There was a tendency of higher error in sloping aquifer than the horizontal aquifer although no definite trend was observed.

The average errors in the unsteady drainage solutions considered overall the three cases of replenishment patterns which were experimentally investigated were 4.43% at midpoint, 6.85% near the lower drain and 8.47% near the drain located on higher side.

The errors were within the reasonable limits for practical use of the analytical solutions.

References

Sewa Ram, 1984. Subsurface drainage of a phreatic aquifer receiving time varying replenishment resting over a sloping impervious barrier. Ph. D. thesis submitted to G.B. Pant University of Agriculture and Technology, Pantnagar, Distt. Nainital (UP), 166 pp.

Sewa Ram and H.S. Chauhan. 1987. Drainage of sloping lands with constant relenishment. Journal of Irrigation and Drainage Engineering, ASCE, 13(3): 212-223.

Sewa Ram and H.S. Chauhan. 1987. Analytical and experimental solutions for drainage of sloping lands with time varying recharge. Water Resources Research. American Geophysical Union, 23(6): 1090-1096.

Problems

Theoretical

1. Draw a neat sketch for the flow of water between two drains/ditches resting over a sloping impervious layer. Write down governing flow

equation for this case and illustrate symbols used in the equations.

2. Write down various possible recharge combinations for which analytical and numercial solutions could be worked out for practical applications.
3. What are the general initial and boundary conditions for solving the governing equation of water flow for a slopy phreatic aquifer? What these conditions physically mean?
4. Derive a steady state solution from the governing equations for slopy phreatic aquifer and transform the solution for non-slopy conditions too.
5. What are the different transformations used for transforming the governing water flow equation to a standard heat flow equations under different recharge patterns?
6. Transform the governing equation and initial and boundary conditions to general heat flow equation using suitable transformation for constant replenishment.
7. Transform the governing equation and initial and boundary conditions to general heat flow equation using suitable transformation for linearly increasing recharge rate.
8. Transform the governing equation and initial and boundary conditions to general heat flow equation using suitable transformation for exponentially declining recharge rate.
9. Transform the governing equation and initial and boundary conditions to general heat flow equation using suitable transformation for a product of linearly increasing and exponentially declining replenishment rate.

Index